MUSIC THEORY TRANSLATION SERIES

Claude V. Palisca, *Editor*

Other Published Volumes in the Music Theory Translation Series:

The Practical Harmonist at the Harpsichord, 1708, by Francesco Gasparini, translated by Frank S. Stillings, edited by David L. Burrows.

The Art of Counterpoint, Part Three of *Le istitutioni harmoniche,* 1558, by Gioseffo Zarlino, translated by Guy A. Marco and Claude V. Palisca.

Hucbald, Guido, and John on Music: Three Medieval Treatises, translated by Warren Babb, edited with introductions by Claude V. Palisca.

The Art of Strict Musical Composition, 1771–1779, by Johann Philipp Kirnberger, translated by David Beach and Jurgen Thym.

Introductory Essay on Composition, 1782–1793, by Heinrich Christoph Koch, translated by Nancy Kovaleff Baker.

On Music, in Three Books, by Aristides Quintilianus, translated with introduction, commentary, and annotations by Thomas J. Mathiesen.

On the Modes, Part Four of *Le Istitutioni harmoniche,* 1558, by Gioseffo Zarlino, translated by Vered Cohen, edited with an introduction by Claude V. Palisca.

The Florentine Camerata: Documentary Studies and Translations by Claude V. Palisca.

Fundamentals of Music by Anicius Manlius Severinus Boethius, translated with introduction and notes by Calvin M. Bower.

Musical Poetics by Joachim Burmeister, translated with introduction and notes by Benito V. Rivera.

THE THEORY OF MUSIC

FRANCHINO GAFFURIO

Translated, with Introduction and Notes,

by **Walter Kurt Kreyszig**

Edited by **Claude V. Palisca**

Yale University Press New Haven and London

The preparation of this work was made possible in part by a grant from the National Endowment for the Humanities, an independent federal agency.

Designed by Sonia L. Scanlon

Set in Bembo type by Keystone Typesetting, Inc., Orwigsburg, Pennsylvania.

Printed in the United States of America by BookCrafters, Inc., Chelsea, Michigan.

Library of Congress Cataloging-in-Publication Data

Gaffurius, Franchinus, 1451–1522.

[Theorica musicae. English]

The theory of music / Franchino Gaffurio ; translated, with introduction and notes, by Walter Kurt Kreyszig ; edited by Claude V. Palisca.

p. cm. — (Music theory translation series)

Includes bibliographical references and index.

ISBN 0–300–05497–1

1. Music—Theory—To 500—Early works to 1800. 2. Music, Greek and Roman—Early works to 1800. 3. Music—Acoustics and physics—Early works to 1800. I. Kreyszig, Walter Kurt. II. Palisca, Claude V. III. Title. IV. Series.

MT5.5.G2613 1993 92-33709

781—dc20 CIP

MN

A catalogue record for this book is available from the British Library.

The paper in this book meets the guidelines for permanence and durability of the Committee on Production Guidelines for Book Longevity of the Council on Library Resources.

10 9 8 7 6 5 4 3 2 1

To Claude V. Palisca

CONTENTS

BOOK THREE

BOOK FOUR

BOOK FIVE

PREFACE BY THE SERIES EDITOR

Few theoretical treatises changed how musicians speculated about music as much as Franchino Gaffurio's *Theorica musice*. Five hundred years ago, when every educated musician had heard of Boethius but very few read him, Gaffurio through the new medium of the printed book placed his interpretations of Boethius and other ancient authors in the hands of hundreds of musicians and scholars. (More than fifty copies of the original printing of 1492 still exist today, a large number for an incunabulum.) In the 1470s and 1480s, when Gaffurio was writing, Boethius was known only through manuscripts—the *Arithmetica* was printed in Augsburg in 1488, and the *Musica* only in 1492 in Venice, the same year as Gaffurio's *Theorica*. Gaffurio had anticipated both with his earlier compendium of Boethius, printed in 1480, the *Theoricum opus musice discipline,* on which the *Theorica* was based.

Just as the young Boethius had made a compendium in Latin of the best sources on music theory available to him—almost all of them in Greek—the young Gaffurio embarked on a similar task, convinced as were so many of his age that the only true learning came from ancient Greece and Rome. Johannes Gallicus had taught him that Boethius was not an original writer theorizing about the music of his time but a transmitter of an enduring body of Greek thought. Because Gaffurio could not read Greek, the Latin Boethius was his first main source; he did not neglect other ancient and more recent Latin authors, such as Cassiodorus, Marchetto, Ugolino, and Anselmi. Only later did he learn of surviving Greek treatises, which he asked other scholars to translate for him.

Whereas Boethius was not a musician, Gaffurio was a singer, composer, teacher of church singers, and from 1484 choirmaster of the Duomo of Milan. Thus in him ancient theory passed through a filter sensitive to its relevance for modern musicians. For unlike Boethius, who was preparing himself and his reader for the study of the liberal arts, Gaffurio was addressing the musician aspiring to be both *cantor* and *musicus,* the latter being a title Boethius bestowed only on the musician able to engage in discourse about his art as well as practice it.

Gaffurio's practical and modern, though conservative, outlook was both an advantage and a weakness. Although he saw how ancient theory could illuminate the modern, he could not shake off the habits of thought of a musician trained in the plainchant and polyphonic tradition. The filter both narrowed and clouded his visions, and he misunderstood substantial parts of what he extracted from Boethius, pseudo-Plutarch, Bacchius, Cleonides, Euclid, and the ancient philosophers. He was also quite uncritical in repeating some of their less well-founded dogmas.

However imperfectly Gaffurio reviewed the theoretical heritage of the past, his compilation served as a foundation for much of the musical speculation that followed. Not always acknowledged, the *Theorica,* together with his *De harmonia musicorum instrumentorum opus* (completed by 1500 but not published until 1518), were the base on which Spataro, Glarean, Zarlino, Salinas, and others built their more original and sometimes deliberately opposing doctrines.

Gaffurio was the last of these authors to use the scholastic method—which was also Boethius's—of almost verbatim transcription, paraphrase, abstract, and interpolated commentary on older texts. Reading Gaffurio's book is reading over his shoulder. But what his contemporaries could do only with great difficulty—get a closer look at what he was reading—this translation makes easy for us, because Walter Kreyszig has punctiliously identified Gaffurio's subtexts by chapter and verse, and many of them are in every good research library. The largest proportion, indeed, are to be found in one of the most recent volumes of this Series: Boethius, *Fundamentals of Music,* Calvin M. Bower's translation of the *De institutione musica.* There is very little that Professor Kreyszig has not succeeded in tracing to some previous, usually ancient, author. Gaffurio may not have read them directly—at the time he wrote *Theorica musice* he evidently had not seen even in translation the musical treatises of Aristoxenus, Nicomachus, Ptolemy, Aristides Quintilianus, the Bellermann anonymi, Porphyry, or Bryennius. Resonances of some of these authors nevertheless abound in the designated Boethian as well as non-Boethian passages indicated in the footnotes.

The preparation of this work was made possible in part by a grant from the National Endowment for the Humanities, for which we are ever grateful.

Claude V. Palisca

TRANSLATOR'S PREFACE

Research on Franchino Gaffurio (1451–1522) has been partial to two of his three major treatises, the *Practica musicae* (1496) and the *De harmonia musicorum instrumentorum opus* (1518).[1] Gaffurio laid the theoretical foundation for these works, however, in his *Theorica musice* of 1492. It covered an array of topics that takes us through the Boethian curriculum and beyond, including the systems of classifying music (Book 1); physical properties of sound and the Pythagorean mathematical ratios (Books 2 and 3), musical intervals (Book 4); and the Greek *systema teleion* and the Guidonian solmization (Book 5). Its numerous sources span nearly two thousand years, from the fourth century B.C.—Plato and Aristotle—to the fifteenth century—for example, Johannes Gallicus and Ugolino of Orvieto. Gaffurio was indebted to Latin authors (notably Boethius, Augustine, Isidore of Seville, Martianus Capella, Giorgio Anselmi, Johannes Gallicus, and Guido of Arezzo) as well as Greek authors (in particular Nicomachus, Ptolemy, Aristotle, Plato, and Bacchius), whose writings were accessible to him through Latin translations and commentaries.

The English translation of Gaffurio's *Theorica musice* is essentially that in my doctoral dissertation, "Franchino Gaffurio's *Theorica musice* (1492): Edition, Translation, and Study of Sources," completed at Yale University in 1989 under the guidance of Professor Claude V. Palisca, which in turn grew out of a project in Professor Palisca's graduate seminar on music theory and aesthetics of the Renaissance in the spring of 1981. I am especially grateful to him for his unfailing wisdom, patient advice, kind criticism, steadfast support, and encouragement. In revising the translation in my doctoral dissertation I benefited also from the careful sympathetic reading and kind suggestions of Professors Reinhard Strohm (King's College, University of London), Craig Wright (Yale University), and Mary Wolinski (Arizona State University, Tempe). Dr. Otto Steinmayer (Department of Classics, Yale University) kindly read the translation and offered expert advice on the idiosyncratic prose style of Gaffurio. Professor Thomas

1. For bibliographies, see Clement A. Miller, "Gaffurius, [Gafurius, Gafori], Franchinus," *The New Grove Dictionary of Music and Musicians,* ed. Stanley Sadie (London: Macmillan, 1980), 7:79; Claudio Sartori, "Gaffurius (Gafori), Franchinus," *Die Musik in Geschichte und Gegenwart,* ed. Friedrich Blume (Kassel: Bärenreiter, 1949–73), 4 (1955), cols. 1242–1243.

Mathiesen (School of Music, Indiana University) read the entire manuscript and made many invaluable observations on Gaffurio's more subtle handling of the Latin language as well as on the identification of *loci paralleli* for a number of passages cited in the *Theorica musice*. The late Stephen Galambos (Windsor, Ontario) gave me a thorough introduction to humanist Latin, characterized by the substantially more complex syntax than that common in medieval Latin texts. I am also grateful for advice and support from a number of other professors and colleagues: Professors Allen Forte (Yale University), Donald Baronowski (University of Saskatchewan, Saskatoon), Michael Bernhard (Bayerische Akademie der Wissenschaften, Munich), Bonnie J. Blackburn (Oxford), Martin Czernin (Schottenstift, Vienna), Charles Fantazzi (University of Windsor), Jan W. Herlinger (Louisiana State University, Baton Rouge), Leofranc A. Holford-Strevens (Oxford), Walter Pass (Institut für Musikwissenschaft, Universität Wien), Heinz Ristory (Chorherrenstift, Klosterneuburg, Austria), John Rothgeb (State University of New York at Binghamton), and Peter Slemon (University of British Columbia, Vancouver). I also wish to thank Professor Harold E. Samuel (Librarian, the John Herrick Jackson Music Library, Yale University) and Mrs. Patricia Howell (Beinecke Rare Book and Manuscript Library, Yale University) for providing easy access to sources.

I am also grateful for generous grants from the Publications Fund of the University of Saskatchewan, awarded by Dr. Michael Owen, Director of Research Services, and from the National Endowment for the Humanities, Washington, D.C., both of which supported the preparation of this manuscript for publication.

Mrs. Yvonne Dopson (Ives Business Services, Saskatoon, Canada) entered the translation into the computer; Mrs. Valerie Ziegler (Saskatoon), a computer graphic specialist, prepared the illustrative material. I wish to thank them for their assistance. I am also indebted to Noreen O'Connor of Yale University Press for her skillful and sympathetic editing of the manuscript.

Finally, I wish to express my sincere gratitude to my parents, Professor Erwin Kreyszig and Dr. Herta Kreyszig, and to my brother, Mag. Herbert Kreyszig, for their constant inspiration and moral support.

Saskatoon, Canada, and Vienna, Austria
January 1993

INTRODUCTION

I collected those things for you [the reader], that you asked for, especially extracting practically all the theory from the music treatise of the excellent Franchino Gaffurio of Lodi. This [theory] seems to me very fruitful and authentic. For he applied his mind to this discipline from his youth on and he became an outstanding professor of the two kinds of music [i.e. theory and practice]. He taught theoretical music and opened up its more secret aspects.[1]

The German music theorist Dietrich Tzwyvel (before 1490–after 1516) here recognized the awesome responsibility undertaken by Franchino Gaffurio (1451–1522) to transmit the musical thought of Greek antiquity.

Gaffurio's theoretical writing, an activity which spanned a large portion of his life, from 1474 to 1521,[2] was intimately connected with his appointments at court,[3] church,[4] and gymnasium[5] but was also greatly influenced by acquaintances with scholars, artists,[6] and poets.

1. "Collegi namque ea tibi, quae petiisti, maxime et excellentis Franchini Gafuri Laudensis musica omne paene sequestrans theoricam, quam illi operi suo immiscet. Haec nobis magis videtur fructifera atque autentica. Ipse enim animum ab ineunte aetate ad hanc disciplinam appulit et utriusque musicae professor eximius evasit atque musicam theoricam edocuit et eius veluti penetralis secretiora aperuit." Dietrich Tzwyvel, *Introductorium musicae practicae,* p. 2, as cited in: Wilfried Kaiser, *Dietrich Tzwyvel und sein Musiktraktat: Introductorium musicae practicae (Münster 1513),* vol. 2 of *Beiträge zur rheinischen Musikgeschichte,* ed. Arbeitsgemeinschaft für rheinische Musikgeschichte (Cologne: Arno Volk, 1956).
2. For a summary of Gaffurio's endeavor in the field of music theory, see Clement A. Miller, "Early Gaffuriana: New Answers to Old Questions," *The Musical Quarterly* 56 (1970):367–368; Irwin Young, "Franchinus Gaffurius: Renaissance Theorist and Composer (1451–1522)," Ph.D. diss., University of Southern California, 1954, pp. 6ff.; Walter Kurt Kreyszig, *Franchino Gaffurio's Theorica musice (1492): A Study of the Sources,* vol. 1 of *Musica Mediaevalis Europae Occidentalis,* ed. Walter Pass (Vienna and Stuttgart: Wilhelm Braumüller, 1993), appendix 2.1.
3. See note 12 below.
4. On January 22, 1484, Gaffurio was appointed *maestro di capella* at the cathedral of Milan, where he was in charge of training choirboys. Cf. Sartori, "Gaffurius (Gafori), Franchinus," col. 1238.
5. Beginning in 1492, Gaffurio lectured on music at a gymnasium in Milan that had been instituted by Lodovico Sforza. Cf. Miller, "Gaffurius, [Gafurius, Gafori], Franchinus," p. 77.
6. For example, Gaffurio's association with Leonardo da Vinci (1452–1519) is discussed in Emanuel Winternitz, *Leonardo da Vinci as a Musician* (New Haven and London: Yale University Press, 1982), pp. 3–9.

Throughout his life, Gaffurio cherished a close friendship with the Milanese poet Lancinus Curtius, who excelled by his erudition in Latin and Greek.[7] Curtius also contributed panegyric poems for the *Theorica musice,* the *Theoricum opus musice discipline,* and the *De harmonia musicorum instrumentorum opus.*[8]

Gaffurio first realized his desire to revive ancient musical learning in his *Theorica musice,* which appeared in two versions. The earlier *Theoricum opus musice discipline*—the first printed book on music theory—was published by Franciscus de Dino Florentinus at Naples in 1480.[9] The volume was dedicated to John Arcimboldi, cardinal bishop of Novara.[10] In the twelve years that elapsed between the *Theoricum opus* and the *Theorica musice,* two important events in Gaffurio's life, namely, his appointment as *maestro di capella* at the Cathedral of Milan in 1484 and his subsequent close association with the court of the Sforzas, paved the way for his more direct contact with the participants and resources of the humanist movement[11] and the rich musical life of the ducal establishment.[12] During this period, Gaffurio was thus able to deepen his knowledge of ancient and medieval music theory. Newly available documents (for example, the unidentified Latin translation of Plutarch's *De musica*)[13] prompted the revision of the *Theoricum opus.*[14] The fruits of these studies were collected in the *Theorica musice,* which appeared in Milan, the first center of a printing league,[15] in 1492. The volume, dedicated to Lodovico

7. Mario Emilio Cosenza, ed., *Biographical and Bibliographical Dictionary of the Italian Humanists and of the World of Classical Scholarship in Italy, 1300–1800* (Boston: G. K. Hall, 1962–1967), 2 (1962):1163–1164.

8. More details on their collaboration are supplied in Clement A. Miller, trans., *Franchinus Gaffurius: De harmonia musicorum instrumentorum opus,* vol. 33 of *Musicological Studies and Documents,* ed. Armen Carapetyan (American Institute of Musicology, 1977), pp. 15–16.

9. Claudio Sartori, *Dizionario degli editori musicali italiani (Tipografi, incisori, librai-editori),* vol. 32 of *Biblioteca di Bibliografia Italiana* (Supplementi a La Bibliofilia), ed. R. Ridolfi (Florence: Leo S. Olschki, 1958), p. 69.

10. The town is twenty-seven miles west of Milan. Cf. *A Concise Encyclopaedia of the Italian Renaissance,* ed. J. R. Hale (London: Thames and Hudson, 1981), p. 226.

11. Elisabeth Pellegrin, "Bibliothèques d'humanistes lombards de la cour des Visconti Sforza," *Bibliothèque d'Humanisme et Renaissance* 17 (1955):218–245; Elisabeth Pellegrin, *La Bibliothèque des Visconti et des Sforzas ducs de Milan, au XVe siècle,* vol. 5 of *Publications de l'Institut de Recherche et d'Histoire des Textes* (Paris: C.N.R.S., 1955).

12. Gaetano Cesari, "Musica et musicisti alla corte sforzesca," *Rivista Musicale Italiana* 29 (1922):3–53; also in: *La Corte di Lodovico Il Moro,* ed. Malaguzzi Valeri (Milan, 1913–1923), 4:183–254; Claudio Sartori, *La Cappella musicale del Duomo di Milano: Catalogo delle musiche dell'Archivio* (Milan: A Cura della Ven. Fabricca del Duomo, 1957).

13. The first published Latin translation by Carlo Valgulio appeared only in 1507, fifteen years after the publication of the *Theorica musice.* Cf. Claude V. Palisca, *Humanism in Italian Renaissance Musical Thought* (New Haven: Yale University Press, 1985), pp. 449–450.

14. For more detailed coverage of these revisions, which include both omissions of previous materials and extension of ideas of the original treatise, see Kreyszig, *Franchino Gaffurio's Theorica musice,* chapter 10.

15. G. Fumagalli, *Lexicon typographicum Italiae: Dictionnaire géographique d'Italie pour servir à l'histoire de l'imprimerie dans ce pays* (Florence: Leo S. Olschki, 1903), p. 213.

Il Moro, duke of Milan,[16] was printed by the well-established Filippo Mantegazza,[17] with the cost of publication borne by Giovanni Petri de Lomazzo.[18] The *Theorica musice* also owed much of its genesis to Jacopo Antiquario, the secretary to the governor of Milan.[19]

In the *Theorica musice,* Gaffurio strikes a balance between the compilation of documents, paraphrase of information, summary of facts, and his own critical remarks. The model for such a successful blending of material is provided in the *De institutione musica* of Boethius—a work which in turn is thought to borrow heavily from Nicomachus's lost major work on harmonics and his *Introduction to Arithmetic.*[20] Gaffurio not only adheres closely to Boethius's overall format and content—the *Theorica musice* like Boethius's *De institutione musica* comprises five books—but, following Boethius, he views music as a science, the science of harmonics. Boethius's *De institutione musica* and *De institutione arithmetica* furnish approximately seventy percent of the content of the *Theorica musice,* the core that is based on ancient Greek harmonic theory, much of it attributable to Nicomachus. In addition to Nicomachus, a number of other Greek writers reached Gaffurio through Boethius, among them, Aristoxenus (*Theorica* 2.2.109; 2.4.25 ff.), Eubulides and Hippasus (*Theorica* 4.5.19–25), Philolaus (*Theorica* 4.3.14) and Ptolemy (*Theorica* 2.2.129 ff., 2.5.23 ff., 5.4.61, 5.8.39), as well as the elusive Latin author Albinus (*Theorica* 2.2.116–119). Gaffurio's *Theorica musice* must therefore be considered an important stage in the retransmission and reinterpretation of Greek theory from Boethius to the late fifteenth and early sixteenth centuries.

Aside from Boethius, Gaffurio consulted many other writers in the preparation of the *Theorica musice.* These include Latin writers: the Fathers of the Church (Augustine, Ambrose), classical poets (Horace, Livy, Ovid, Virgil), and the encyclopaedic authors Isidore of Seville and Martianus Capella—two

16. Gaffurio also dedicated his *Practica musicae* to Lodovico Sforza.

17. Sartori, *Dizionario degli editori musicali italiani,* p. 94; Cosenza, *Biographical and Bibliographical Dictionary of Italian Printers and of Foreign Printers in Italy from the Introduction of the Art of Printing in Italy to 1800* (Boston: G. K. Hall, 1968), p. 383.

18. Lomazzo also lent financial support to the publication of Gaffurio's *Practica musicae.*

19. *Theorica,* Dedication, 27. Little is known about the humanist Jacopo Antiquario of Perugia (1444?–1512?). Antiquario's *Oratio pro populo,* an oration addressed to Louis XII of France in the name of the Milanese populace, documents Antiquario's position as secretary to the duke of Milan as well as his involvement in important diplomatic missions. Upon the seizure of Milan by Louis XII, Antiquario went over to the conqueror's side. I thank Professor Palisca for drawing my attention to this significant document in the Beinecke Rare Book and Manuscript Library of Yale University. Also, in a letter of 1 September 1479 to Lodovico Maria Sforza, Jacopo Antiquario pleads for a benefice for Gaffurio. Cf. Palisca, *Humanism,* p. 8.

20. On Boethius's method of compiling through translation, which is interspersed with his own commentary, see Calvin M. Bower, "Boethius and Nicomachus: An Essay Concerning the Sources of De institutione musica," *Vivarium* 16 (1978):1–45.

authorities whose writings survived in copious manuscripts and printed editions.[21] As Gaffurio lacked a reading knowledge of Greek, he followed the example of his uncle Taddeo Fissiraga, who once had a treatise of Plutarch translated,[22] and engaged several humanists to translate Greek texts.[23] Among them was Francesco Burana, who prepared a Latin version (no longer extant) of the musical treatise of Bacchius.[24] In addition, Gaffurio resorted to other Latin translations of previously unknown treatises; in the *Theorica musice* he incorporated material from Marsilio Ficino's Latin translations of and commentaries on Plato, Ermolao Barbaro's Latin version of Themistius's *Paraphrases* on the *De anima* of Aristotle, and Pietro d'Abano's commentaries on Aristotle's *Problems*. Of course, Plato's views were in part available to Gaffurio through both Boethius (though undoubtedly in very limited form)[25] and the lesser known Chalcidius.[26] In the case of the other major treatises of Aristotle, in particular the *Physics,* the Latin mediating source cannot be securely determined. Indeed, the favorable reception of Aristotle during the Middle Ages and the early Renaissance[27] may account for the copious references to Aristotle in the *Theorica musice*. In the case of a few authors, surviving documents shed some light on the question of ownership of a particular volume. For example, marginalia in Gaffurio's own hand are found in his personal copies of Ugolino of Orvieto's *Declaratio musicae disciplinae*[28] and Giorgio Anselmi's *De musica*.[29] Yet in the case of Augustine's *De musica,* one of

21. Charles Henry Beeson, *Die Lagerung und Verbreitung der Handschriften des Isidorus von Sevilla* (Munich: C. H. Beck'sche Verlagsbuchhandlung, 1931); Robert Benaway Brown, *The Printed Works of Isidore of Seville,* vol. 5 of *Occasional Contributions* (University of Kentucky Libraries, 1949); Claudio Leonardi, "I Codici di Marziano Capella," *Aevum* 33 (1959):443–489; 34 (1960):1–524.

22. Gaffurio's uncle Taddeo Fissiraga was a priest, a doctor of science, and also a student of the humanities. Cf. Young, "Franchinus Gaffurius," p. 7.

23. F. Alberto Gallo, "Le traduzioni dal Greco per Franchino Gaffurio," *Acta Musicologica* 35 (1963):172–174.

24. Palisca, *Humanism,* p. 111. Palisca further mentions several other translations commissioned by Gaffurio. These include: Ptolemy's *Harmonics,* Aristides Quintilianus's *De musica,* Bryennius's *Harmonics,* and three anonymous manuals (known until recently as Bellermann's anonymi) (Palisca, *Humanism,* p. 191)—all of which must have been completed after 1492 as they presumably were not used in the preparation of the *Theorica musice.*

25. Boethius, *De inst. mus.* 1.30, Calvin M. Bower, trans., *Anicius Manlius Severinus Boethius: Fundamentals of Music,* ed. Claude V. Palisca (New Haven and London: Yale University Press, 1989), p. 48.

26. *Timaeus a Calcidio translatus commentarioque instructus,* ed. J.H. Waszink, vol. 4 of *Plato Latinus,* ed. Raymund Klibansky as part of *Corpus Platonicum Medii Aevi* (London: Warburg Institute, and Leiden: E. J. Brill, 1962).

27. Clemens Bäumker, "Zur Rezeption des Aristoteles im lateinischen Mittelalter," *Philosophisches Jahrbuch der Görresgesellschaft* 27 (1914):478–487.

28. Albert Seay, "Ugolino of Orvieto, Theorist and Composer," *Musica Disciplina* 9 (1955): 112.

29. Jacques Handschin, "Anselmi's Treatise On Music Annotated by Gafori," *Musica Disciplina* 2 (1948):123–140.

the works most widely disseminated across Europe,[30] the copy from which Gaffurio worked, if indeed he owned one, cannot be determined. An inventory of Gaffurio's library testifies to a rather extensive collection, including several books cited in the *Theorica musice,* among them, Guido of Arezzo's *De musica* [i.e., *Micrologus*], Boethius's *De arithmetica,* and Marchetto of Padua's *De musica* [i.e., *Lucidarium*]).[31]

Apart from close contacts with the court of the Sforzas in Milan, Gaffurio also had associations with various other establishments. For example, his visit at the court of Lodovico Gonzaga at Mantua from 1474 to 1476[32] resulted in his becoming familiar with Johannes Gallicus's *Libelli musicalis de ritu canendi vetustissimo et novo*[33]—a work which Gaffurio cites at great length in the *Theorica musice* (5.4). In addition to original treatises in Latin and Latin translations of Greek texts, Gaffurio on occasion also resorted to correspondence between authors, for example, the letter of Cassiodorus to Boethius (*Theorica* 5.8.46).[34] Thus Gaffurio tapped many of the commonly known transmitters of universal knowledge during the Middle Ages and the Renaissance: compendia, philosophical commentaries, treatises, letters, and translations.[35] Also, personal friendships provided additional inspiration for his writing. Most notable is Gaffurio's contact with Johannes Tinctoris (c. 1435–1511?), who clarified arithmetical and musical terms in his *Proportionale musices* (1473–1474)[36] and in his *Terminorum musicae diffinitorium* (1495).[37]

30. Manfred Oberleitner, *Die handschriftliche Überlieferung der Werke des Heiligen Augustinus Band I/1: Werkverzeichnis,* vol. 1 of *Veröffentlichungen der Kommission zur Herausgabe des Corpus der lateinischen Kirchenväter,* ed. Rudolf Hanslick as vol. 263 of *Österreichische Akademie der Wissenschaften Philosophisch-Historische Klasse: Sitzungsberichte* (Vienna: Hermann Böhlaus, 1969):1–407, also vols. 267 (1970):1–384; 276 (1972):1–340; and 281 (1972):1–394.

31. A more complete inventory is given in Alessandro Caretta, "La biografia di F. Gaffurio nel Cod. Lauden. XXVIII. A.9," in Alessandro Caretta et al., *Franchino Gaffurio* (Lodi: Edizioni Dell'Archivio Storico Lodigiano, 1951), pp. 117–118.

32. Young, "Franchinus Gaffurius," p. 15.

33. Johannes Gallicus (c. 1415–1473) was a pupil of Vittorino da Feltre. Cf. Cecil Adkins, "Legrense [Carthusensis, Gallicus, Mantuanus], Johannes [Jean de Chartreaux; Jean de Namour]," *New Grove* 19:614. Da Feltre founded a school whose curriculum included music as a humanist subject at the court of Gianfrancesco Gonzaga in Mantua in 1424. Cf. Palisca, *Humanism,* p. 7.

34. For the content of this letter, see Günther Wille, *Musica Romana: Die Bedeutung der Musik im Leben der Römer* (Amsterdam: P. Schippers, 1967), pp. 700–704.

35. Paul Oskar Kristeller, "The Scholar and His Public in the Late Middle Ages and the Renaissance," in *Medieval Aspects of Renaissance Learning: Three Essays by Paul Oskar Kristeller,* ed. and trans. Edward P. Mahoney, vol. 1 of *Duke Monographs in Medieval and Renaissance Studies* (Durham, North Carolina: Duke University Press, 1974), pp. 3–25.

36. For a translation of this treatise, see Albert Seay, "The *Proportionale Musices* of Johannes Tinctoris," *Journal of Music Theory* 1 (1957):22–75.

37. For a facsimile edition with a German translation, see *Johannes Tinctoris: Terminorum musicae diffinitorium* (Facsimile of Incunable, Treviso, 1495), ed. Heinrich Bellermann, vol. 37 of *Documenta musicologica, Series I: Druckschriften-Faksimiles,* ed. Association Internationale des Bibliothèques Musicales (Kassel: Bärenreiter, 1983).

Book **One**

The obvious influence of many diverse currents of thought and underlying philosophies, mainly of previous eras, appear from the beginning of the *Theorica musice*. The opening chapter of the treatise, a testimony to Gaffurio's skill in selecting from abundant material, his capability to synthesize information at hand, and his expertise in presenting ideas and thoughts in a coherent and original fashion, examines the role music played in the societies of Greece and Rome during antiquity and the Middle Ages. Gaffurio's synopsis of Greek *mousike*—a term embracing the entire body of concepts and ideals spanning from the evolution of musical instruments, the significance of dance, concepts of *melos, harmonia, rhythmos,* modality, to the broader topics of music aesthetics and music education—is ingeniously cast within an overview of significant contributors, including both practicing musicians (such as the kitharode Arion) and authors engaged in the writing of treatises (for example, Isidore of Seville). In order to reflect the vastness of this tradition in the relatively small space allotted for that compact survey, Gaffurio returns to various topics and authors (briefly cited in the opening chapter) later in the *Theorica musice*. For example, at the opening of Chapter 1, in the context of a brief inquiry into the origin of music, he ponders the vexing issue concerning the inventor of music—a position for which two contenders are considered: the biblical Jubal, from the tribe of Cain, whose invention of music was to delight the shepherds,[38] and Pythagoras, who is credited with the discovery of numerical laws of musical proportions through experimentation with hammers.[39] Gaffurio resolves this issue only in the final book of the *Theorica musice* (5.1.4), when Jubal ultimately emerges as the inventor of music.[40]

Gaffurio fuses Christian theology and Greek mythology in the rather lengthy citation from Giorgio Anselmi's *De musica* concerning the hierarchical order of angels (*Theorica* 1.1.165–194).[41] The close affinity between music and

38. For a comprehensive discussion, see Judith Cohen, "Jubal in the Middle Ages," *Yuval* 3 (1974):83–99.

39. For a thorough historical survey, see James McKinnon, "Juval vel Pythagoras, qui sit inventor musicae?" *The Musical Quarterly* 64 (1978):1–28.

40. Given Gaffurio's position as a staunch supporter of the Pythagorean-Platonic tradition, such resolution of the dilemma clearly attests to his own roots in the biblical tradition. Even Johannes Gallicus, a fervent defender of Boethius's theories, ascribes the origin of music not to Pythagoras but to Jubal. Cf. Johannes Gallicus, *Ritus canendi vetustissimus et novus* 1.1, in *Scriptorum de musica medii aevi novam seriem a Gerbertiana alteram,* ed. Edmond de Coussemaker, 4 vols. (1864–1876; reprint Hildesheim: Georg Olms, 1963), 4:299–300; a more recent edition is *Joannes Gallicus Legrense: Ritus canendi vetustissimus et novus,* ed. Albert Seay, *Colorado College Music Series,* ed. Albert Seay (Colorado Springs: Colorado College Music Press, 1981).

41. Cf. , *Giorgio Anselmi Parmensis, De musica,* ed. Giuseppe Massera, vol. 14 of *Historiae Musicae Cultores Biblioteca* (Florence: Leo S. Olschki, 1961).

religion is succinctly illustrated in the many musical forms of celebration and veneration of gods in hymns, psalms, and so on, with recourse to both instruments and voices (*Theorica* 1.1.155 ff., 289, 292).[42] Proof of the inherent force of music is also documented in several other accounts, for example, in David's playing of the lyre effecting the expulsion of the evil spirits from King Saul (*Theorica* 1.1.75–76) and Arion's playing of the kithara which leads to his rescue by the dolphin (*Theorica* 1.1.65). Gaffurio devotes considerable attention to the Platonic doctrine of ethos and the concept of *paideia.* He also discusses music in the context of rhetoric and medicine.

Gaffurio roots *musica mundana,* the orderly revolution of the world, firmly in mathematics. The resultant system of astronomy provides a basis for identifying certain motions of the spheres with the tones of a musical scale. Arithmetic also dominates the interpretation of *musica humana.* The formulation of the ancient doctrine of ethos lies at the heart of this concept, which is based on the mutual dependency of body and soul controlled by harmonious numerical proportions. Gaffurio surveys some prudently chosen topics (*Theorica* 1.3): the natural relationship of the human soul with *harmonia*; the pulse theory of Hierophilus; the Platonist's description of the soul as *harmonia,* communicated explicitly in the writings of Philolaus; and the philosophical theories of embryology, transmitted in Censorinus's *De die natali.*

With respect to *musica mundana* and *musica humana* the emphasis is placed squarely on the rational interpretation of music, seeking out proportions that provide the link between music and cosmos in the case of *musica mundana,* and between music, the human soul, and the body in the case of *musica humana*; the discussion of the third category, namely, that of *musica instrumentalis,* concerns the timbral classification of music according to the object responsible for emitting the sound (*Theorica* 1.4), either by natural instruments, artificial instruments (*musica organica,* according to Augustine and Gaffurio), rhythmical instruments (*musica rhythmica,* according to Augustine and Gaffurio), or by a combination of the latter two categories (*ex utriusque composito,* according to Gaffurio). The reporting (in *Theorica* 1.4) of the well-known dichotomy of *musicus,* who is knowledgeable in music theory and comprehends what he performs, and *cantor* or *phonascus,* who performs without the knowledge of theory, attests to Gaffurio's ardent belief in the healthy interaction between theory and practice in the consummate musician. Augustine's threefold division of the living being, namely, *musicus, cantor,* and *animal* (*Theorica* 1.6.14–19), has a correlation in the terms *scientia, ars,* and *natura*—the latter two branches being described by Gaffurio as "divine" (*Theorica* 1.6.21), that is,

42. For a comprehensive discussion of this and other topics, see Kreyszig, *Franchino Gaffurio's Theorica musice.*

considered beyond human comprehension and Gaffurio exhibits an inexplicable rationality. On the other hand, Gaffurio shows an obvious bias toward Pythagorean thought in his reliance on numbers over and against sense-discrimination. Boethius is quoted as saying, "It is possible not to perceive very small things because of the minuteness of the sensible objects themselves, and sense perception is frequently confused by very large objects."[43] This justifies Gaffurio's reporting of the Pythagorean discovery of the consonances at a blacksmith's shop (*Theorica* 1.8), a myth that enjoyed well over a thousand years of continuous transmission.[44] In spite of its grave fallacies in the observation of natural phenomena, it is recounted here not only in the prose of Boethius's *De institutione musica* but is also captured graphically in the woodcut at the close of Chapter 8 of Book 1.

Book **Two**

Book 2 of the *Theorica musice* is devoted to an examination of the acoustical properties of sound and to the mathematical foundations of proportions. The desire to bring the Greek tradition into closer affiliation with the Scriptures, characteristic of the opening chapter of the *Theorica musice,* leads Gaffurio to an incorrect etymology for the word *musica*. Gaffurio considers the elements of music, namely the diastema (tones and semitones) of the *systema teleion* of the Greeks, in preparation for his exposition of the theories of motion, sound, and its oral perception, all by way of paraphrases from Greek treatises in translation, namely, Aristotle's *Physics* (the Latin edition consulted by Gaffurio is not identifiable) and Themistius's *Paraphrases* of Aristotle's *De anima* in a translation by Ermolao Barbaro. On occasion, Gaffurio also resorts to Latin treatises, for example, to Boethius's *De institutione musica* for the definition of sound. Gaffurio focuses on various types of motions and corresponding sounds, explaining the relationships between low sounds and slow motions and between high sounds and fast motions, for which he is indebted to Boethius's *De institutione musica*. In these deliberations, Gaffurio keeps separate the Aristotelian and the Pythagorean/Platonic traditions, the latter largely transmitted through Boethius. The introduction of the term *vox* (literally translated as "voice"), referring to the voice of every living being in the classification of Aristotle, as well as in a metaphorical sense to the sounds produced on instruments, as reported in Themistius's *Paraphrases* on the *De*

43. Boethius, *De inst. mus.* 1.9.196. Cf. Bower, trans., *Boethius: Fundamentals,* p. 17.

44. Barbara Münxelhaus, *Pythagoras musicus: Zur Rezeption der pythagoreischen Musiktheorie als quadrivialer Wissenschaft im lateinischen Mittelalter,* vol. 19 of *Orpheus-Schriftenreihe zu Grundfragen der Musik,* ed. Martin Vogel (Bonn and Bad Godesberg: Verlag für Systematische Musikwissenschaft, 1976).

anima, leads Gaffurio to a brief explanation of this terminology in the animal world, considering the emission of sound not "by natural instruments" (to borrow Aristotle's term) but with the aid of a mechanism. This excursion provides a further delineation of the terms *vox* and *sonus,* and thus specifies various conditions necessary for the formation of both *vox* and *sonus* (*Theorica* 2.1.67–69). Boethius's *De institutione musica* is mentioned here to allow for a distinction of various registers of the voice in the terms *acutus* (high) and *gravis* (low) and thus a consideration of the relation between *vox* and *motus* (motion) (*Theorica* 2.1.39–44, 90–94), with the discussion more firmly anchored in arithmetic, as is seen in the use of the terms *pluralitas* (plurality) and *paucitas* (paucity), Boethius's reference to numerical representation of physical phenomena. In connection with materials and conditions necessary for the formation of sound, Gaffurio briefly touches on the topic of resonance and reverberation (*Theorica* 2.2.13–31). For the various categories of voice motions, among them *vox continua* (continuous voice), *unisonae* (unisons), and so forth, Gaffurio is indebted to Boethius, who in turn paraphrased Nicomachus, Ptolemy, and Albinus (*Theorica* 2.2.110–150).

Gaffurio's lack of knowledge of Greek becomes most apparent in his comments on the theories of *rhythmos,* as he was not able to peruse the earliest Greek sources and instead paraphrases Augustine's systematic classification of rhythm in the *De musica* according to one of five genera, that is, *numeri sonantes, occursores, progressores, numeri recordabiles,* and *numeri iudiciales.*

The focus on terminology continues in Chapter 3 of Book 2, where Gaffurio defines the constituent elements of music: diastema, systema, diesis, apotome, tone, and semitone—the information adopted from Bacchius's *Introductio artis musicae* and Boethius's *De institutione musica.* Gaffurio takes a cursory glance at some selected terms in acoustics: *symphonia, melodia,* and *harmonia,* extracted from a broad spectrum of definitions offered by Bacchius, Boethius, Fabius Quintilian, the anonymous *Musica enchiriadis,* and Marchetto of Padua (*Theorica* 2.3.33–41). Insights gained from the examination of some basic principles of sound propagation, including the reflection and diffraction of waves by obstacles, are supplemented by divergent philosophies of Platonists and Stoics (*Theorica* 2.4), differing views of Aristoxenians and Pythagoreans concerning the judgment of consonances, the division of the whole tone (in passing), as well as the atomic theory of sound.

In Chapters 5 through 8 of Book 2, Gaffurio discusses some of the basic properties of proportions and their use in the determination, description, and classification of consonances. On the theory of numbers, Gaffurio follows the footsteps of earlier Latin theorists (Boethius, Isidore of Seville, Ficino, Ambrose, Augustine, and Martianus Capella) on the significance, definition, classification, principle, progression, and etymology of numbers, touching briefly on number symbolism.

Book **Three**

In Book 3, Gaffurio focuses on the concepts of proportion (or ratio) and proportionality (or ratio of ratios). The analogy between discrete sounds, identifiable without ambiguity by assigning numbers, and discrete quantities, delineated by their parts which approach the unit (*Theorica* 3.1), allows Gaffurio to propose a hierarchical ordering of discrete quantities according to multiple, superparticular, and superpartient ratios. Similarly, in the discussion of genera and species of proportions, Gaffurio concentrates only on arithmetic principles relevant to his unfolding of harmonic science, thus quickly dismissing the five genera of smaller inequality and focusing his investigation on the five genera of greater inequality, that is, three simple genera (multiple, superparticular, superpartient) and two composite genera (multiple-superparticular and multiple-superpartient), each of which revolves around the arrangement of two unequal commensurable quantities in various formations (*Theorica* 3.2). The emphasis here is squarely placed on those proportions comprising the first four positive integers (the *tetraktys*) of the Pythagoreans as well as composite proportions, whose terms can be reduced to this basic numerological series. The inclusion of the superpartient ratio (5:3) is justified only in that it provides solid ground for repudiating Ptolemy's divergent classification of the diapason-plus-diatessaron as a consonance.[45] Owing to the complexity of ratios they embrace, the two composite genera definitely have no place in the harmonic science of the Greeks; their inclusion here is seen as Gaffurio's conscious extension of the application of arithmetic proportions to the rhythmic complexities of the mensural system, thereby responding to his earlier call for the fusion of theory and practice in the consummate musician (*Theorica* 1.5). For the information on the genera and species, Gaffurio is primarily indebted to Ugolino of Orvieto, who in his *Declaratio musicae disciplinae* amalgamates relevant material from Boethius's *De institutione musica* and *De institutione arithmetica*. A broad discussion in the *Theorica musice* is devoted to the classification of proportions of greater inequality according to genera (multiple proportion, superparticular proportion, and so on) and species (duple proportion, triple proportion, and so on), the relationship between genera and species of genera through geometric progression, as well as the formation of species of genera through arithmetic progression.[46]

The foregoing classification of proportions according to genera and species of genera, centered around the concepts of equality and inequality, is broadened in Chapter 3 of Book 3 to introduce a new term, proportionality—

45. This topic is also discussed in *Theorica* 2.5.40–58.

46. *Theorica* 3.3–5. For a summary, see Kreyszig, *Franchino Gaffurio's Theorica musice*, tables 7.3, 7.4, and 7.5.

denoting a conjunction of two or more successive proportions that have been arranged from two or more successive terms—and various categories of proportionality, including arithmetic conjunct, arithmetic disjunct, geometric conjunct, geometric disjunct, and harmonic (*Theorica* 3.7.38–57, 66–76). In this context, Gaffurio also mentions the mathematical mean, offering the possibility of modifying the deviation of the middle term of a series—a topic reserved for a more detailed discussion in the *De harmonia*.[47] The thorough examination of the arithmetic proportions—in fact, that topic which receives a more comprehensive coverage than any other aspect discussed in the *Theorica musice*—lays the ground for a consideration of the Greek *harmonia,* and that in the context of Pythagorean arithmetic.

Within the broad spectrum of the term *harmonia*—a term coined by the Greeks, circumscribing both the formation and the principle of order of the entire universe[48]—Gaffurio devotes his examination to the *intervallum,* where the intimate connection between *harmonia* and numerology becomes readily apparent. Gaffurio concentrates on the derivation of the consonances as well as on the whole tone from the perspective of Pythagorean arithmetic and thus is able to sketch the emergence of the Pythagorean scale, firmly rooted in the Greek *tetraktys* (*Theorica* 3.8).

Book **Four**

Book 4, like Book 3, is predominantly mathematical in orientation, as Gaffurio continues with a discussion of the intervals in music by way of the mathematical explanations of multiple and superparticular proportions. In his explanations he consistently adheres to the Pythagorean view, even in the three divergent classification schemes of intervals by Hippasus, Eubulides, and Nicomachus, as well as the standard Pythagorean tuning[49]—all of which dwell on the primacy of the diapason. However, unlike the deliberations in Book 3 of the *Theorica musice,* which concentrate on the *tetraktys* both in its simple form (1, 2, 3, 4) and in its expanded version (6, 8, 9, 12), Gaffurio introduces a new element of the Greek *harmonia,* the *senario*. In Chapter 3, Gaffurio concentrates on the unequal division of the whole tone, which necessitates the introduction of a new set of terms, including the apotome, diesis,

47. *De harm.* 3.1–5. For a synopsis of proportionality and mean, see Kreyszig, *Franchino Gaffurio's Theorica musice,* table 7.7.

48. Thomas J. Mathiesen, "Problems of Terminology in Ancient Greek Theory: 'APMONIA,'" *Festival Essays for Pauline Alderman: A Musicological Tribute,* ed. Burton L. Karson et al. (Provo, Utah: Brigham Young University Press, 1976), pp. 3–17.

49. For a summary of these orderings, all of which are reported in Boethius's *De inst. mus.* 2.19–20, see Kreyszig, *Franchino Gaffurio's Theorica musice,* table 8.6.

comma, schisma, and diaschisma.[50] The numerical calculations of these intervals constitute a rebuttal of Aristoxenus, who insisted on an equal division of the whole tone and equated the diapason with six whole tones—a system which Boethius had reported with disapproval.[51] Gaffurio assembles his information on the numerical derivation of these terms from a variety of sources, among them the followers of the Pythagorean doctrine; a group of later theorists not identified by Boethius (and accordingly also not by Gaffurio), though likely including Theon of Smyrna and Gaudentius; and Philolaus, who displays a special interest in the correlation between comma, apotome, and diesis, expressing their interdependence in terms of mere numbers rather than according to the Pythagorean convention in ratios of numbers. Given the survey of such diverse dissemination and manifold explanations of Pythagoras's doctrine and philosophy, certain irreconcilable notions are bound to emerge. For example, the diesis is termed a smaller semitone by the Pythagoreans,[52] and half of a smaller semitone by later theorists.[53] Gaffurio by no means aims here at a comprehensive coverage of the topic; his rather conservative approach disregards the more recent notions about the division of the whole tone into five equal parts (called diesis) by Marchetto of Padua, the first theorist to propose such a division.[54] Gaffurio's familiarity with Marchetto's *Lucidarium*—the treatise in which the equal division of the whole tone first surfaced (*Lucidarium* 2.5–7)—is evidenced not only in the citations from this treatise in the *Theorica musice* (*Theorica* 2.1.2, 2.3.35, 5.6.100) but also in his having copied the entire *Lucidarium* during 1473–1474.[55] The deliberate omission of the alternative tone/semitone divisions of Marchetto can only be explained by Gaffurio's wish to suppress innovative theories of the division of the whole tone.

Book **Five**

If the preceding four books rely almost exclusively on Greek music theory, the final book of the *Theorica musice* broadens the scope to Latin music theory,

50. For mathematical calculations of these terms, see Kreyszig, *Franchino Gaffurio's Theorica musice,* table 8.4.A.
51. Boethius, *De inst. mus.* 3.1.
52. Boethius, *De inst. mus.* 2.28.260.21; cf. *Theorica* 4.3.9.
53. Boethius, *De inst. mus.* 3.5.277.3–4; cf. *Theorica* 4.3.11.
54. Jan W. Herlinger, "Marchetto's Division of the Whole Tone," *Journal of the American Musicological Society* 34 (1981):193–216.
55. Manuscript Tremezzo, private library of Count Gian Ludovico Sola–Cabiati, manuscript without signature (olim Lodi-Vecchia, Benedictine Abbey of San Pietro), copied by Gaffurio, includes the *Lucidarium* on pages 27–61. Cf. Jan W. Herlinger, *The Lucidarium of Marchetto of Padua: A Critical Edition, Translation, and Commentary* (Chicago: University of Chicago Press, 1985), 46–47.

adding, for example, such topics as the Guidonian solmization and the systems of hexachords, which are oriented towards the practice of music. This fusion of theory and practice not only clarifies and enlivens the rather abstract issues of the Greek *systema teleion* but also affirms Gaffurio's comprehensive view of the discipline.

In Book 5, Gaffurio disassociates himself somewhat from the Pythagorean arithmetic, as is clearly noticeable in his description of the consonances, not in the context of arithmetic ratios but in terms of steps and intervals (*Theorica* 5.1 ff). Gaffurio covers the structural components of the "Perfect Immutable System" (embracing the fifteen-step gamut) but also offers a historic synopsis of the tetrachord's gradual unfolding through Terpander, Hyagnis, and Timotheus of Milet. Gaffurio's lengthy paraphrase of Boethius's *De institutione musica* on the topics of the invention and designation of steps, for which Boethius was probably indebted to Nicomachus's lost *opus maior,* is supplemented by the interpolation of several peripheral sources, namely, Virgil's *Aeneid* and Varro's lost *Hebdomades* (*Theorica* 5.1.14–20). Lack of consensus among theorists about how to arrange the inner pitches of the tetrachord[56] and the scarcity of Greek sources in translation probably made Gaffurio reluctant to place his discussion in the context of arithmetic. For the geometric projection (with the aid of a compass) of the fifteen-step Perfect Immutable System on the monochord (*Theorica* 5.4), Gaffurio is indebted to Boethius, who presented the monochord divisions in the Pythagorean tuning.[57] Gaffurio outlines the fifteen steps of the *systema teleion* in the diatonic genus and its correlation with the planetary orders of Boethius and Cicero (*Theorica* 5.3), thus recalling the earlier discussion of *musica mundana* (*Theorica* 1.2). With the arrangement of whole and half steps in place, Gaffurio clarifies the order of the semitones in the Perfect Immutable System, namely, and through the fusion of diatonic and chromatic genera (*Theorica* 5.5).

Gaffurio paraphrases Guido's *Micrologus,* the most widely disseminated volume on music in the Middle Ages next to Boethius's *De institutione musica,*[58] on the topics of the Guidonian solmization and the system of hexachords—both of which further enhanced the understanding of Greek harmonics. Despite ready access (in Boethius's *De institutione musica*)[59] to the Greek notational symbols for the steps on the monochord, Gaffurio refrains from includ-

56. André Barbera, "Arithmetic and Geometric Divisions of the Tetrachord," *Journal of Music Theory* 21 (1977):294–323.

57. Boethius presumably adopted this information from Euclid's *Division of a Monochord*; for a translation of this document, see Thomas J. Mathiesen, "An Annotated Translation of Euclid's Division of a Monochord," *Journal of Music Theory* 19 (1975):236–258. Although Gaffurio owned a copy of Euclid's treatise (cf. inventory in Caretta et al., *Franchino Gaffurio,* pp. 117–118), his lack of knowledge of Greek made him depend on Boethius.

58. Claude V. Palisca, "Guido of Arezzo," *New Grove* 7:803.

59. Boethius, *De inst. mus.* 4.16.

ing these "difficult ciphers" (except for gamma)[60]; instead, he assigns Latin letters to the corresponding fifteen steps of the monochord, with recourse to the Guidonian system of solmization.[61] Though undoubtedly familiar with the graphic representation of the Guidonian hand, found in numerous treatises of contemporary writers—for example, Johannes Tinctoris's *Expositio manus*[62]—Gaffurio does not resort to this illustration in the *Theorica musice.* The incompleteness of the account of the Latin system is further evident in Gaffurio's partial adoption, from Marchetto's *Lucidarium,* of the classification of hexachords into soft, hard, and natural, which fails to place this terminology in the context of hexachord mutation (*Lucidarium* 8.2).

The Guidonian system provides Gaffurio a platform in the concluding chapter of the *Theorica musice* for a comparison of Greek tonality and Western modality. The discussion of the species of diatessaron, diapente, and diapason in Chapters 7 and 8 of Book 5 serve, in turn, as a springboard for the rather cryptic, yet faulty, remarks on the Western modes. Gaffurio's confusion of octave species, modes, and tonoi[63] was to be compounded in the *De harmonia,* where he tries to reconcile two further treatises in translation, namely, Giorgio Valla's translation of Cleonides's *Introduction to Harmonics* (1497) and Nicolo Leoniceno's of Ptolemy's *Harmonics* (1499).[64]

In view of Gaffurio's defense of the Pythagorean doctrines throughout the *Theorica musice,* the concluding references to the non-Boethian theory of the modes, namely, to the fifteen-mode system of Cassiodorus and Martianus Capella (*Theorica* 5.8.47–57) can only be interpreted as Gaffurio's conscious striving toward a more comprehensive exposition of the topic. Cassiodorus's classification of the fifteen modes according to their low, middle, and high ranges (indicated by the prefixes hypo-, simple, and hyper- forms added to their ethnic names) recalls the issue of the modal ethos—a topic first alluded to in the opening chapter of the *Theorica musice* (*Theorica* 1.1.80 ff.). But the utter confusion concerning Greek tonality and Western modality must lead us to

60. *Theorica* 5.6.11. In his decision, Gaffurio was perhaps influenced by a remark of Vitruvius, who stated that Greek harmonics is "an obscure and difficult subject to read and write about, particular for those who do not know Greek letters." (*De arch.* 5.3, as cited and translated in Claude V. Palisca, "Introductory Notes on the Historiography of the Greek Modes," *Journal of Musicology* 3 (1984):221).

61. Cf. figure 5.6.3 below.

62. Albert Seay, "The Expositio manus of Johannes Tinctoris," *Journal of Music Theory* 9 (1965):194–232.

63. For further details on the misinterpretation of these terms, see Palisca, *Humanism,* pp. 293ff. For a clarification of these terms, see the papers delivered and published as part of the Round Table "The Ancient Greek Harmoniai, Tonoi, and Octave Species in Theory and Practice," Louisville, Ky., October 27, 1983, all published in *The Journal of Musicology* 3 (1984): André Barbera, "Octave Species," 229–241; Calvin M. Bower, "The Modes of Boethius," 252–263; Thomas J. Mathiesen, "Harmoniai and Ethos in Ancient Greek Music," 264–279; and Jon Solomon, "Towards a History of *Tonoi,*" 242–251.

64. Palisca, *Humanism,* pp. 43, 45.

the conclusion that Gaffurio's aspiration "to cleanse the obscure and false" of predecessors and contemporaries (*Theorica,* [Dedication] 18) remained unfulfilled.

That Gaffurio was greatly admired by his contemporaries as well as by music theorists of successive generations can be surmised from the remarks of enthusiastic admirers, such as Carlo Valgulio (the translator of Plutarch's *De musica* into Latin), who extolled Gaffurio as "that excellent musician of our age" whose books "are composed with elegance and erudition,"[65] Johannes Turmair, (1477–1534) who credited Gaffurio among the more recent authors as one "who understands and explains this art eruditely,"[66] and Dietrich Tzwyvel, whose comments were cited earlier. The *Theorica musice* holds an undisputably significant place in the history of late fifteenth- and early sixteenth-century music-theoretical thought, as is readily recognized in the circulation of Gaffurio's ideas in various sixteenth-century treatises, though, as Johannes Turmair notes, "[Many authors] do not give him credit, since they prefer to seize furtively than to admit through whom they have profited."[67] Heinrich Glarean (1488–1563) remains an exception. Book 1 of his *Dodecachordon* (Basel, 1547), which concentrates on the elements of music as well as on topics of consonance, dissonance, and solmization, is derived largely from Boethius and Gaffurio, both of whom are cited as authorities.

The *Theorica musice* came under severe criticism from Giovanni Spataro (1485?–1541), over questions of tuning.[68] Spataro's mentor, Ramos de Pareja (c. 1440–1491 or later), in his attack on Boethius's tuning system, had suggested a representation of the interval of the major third by the simplified ratio

65. Valgulio, *Prohemium* to Plutarch's *De musica* 244 D, as cited in Claude V. Palisca, *The Florentine Camerata: Documentary Studies and Translations* (New Haven and London: Yale University Press, 1989), p. 17.

66. Johannes Turmair, *Musicae rudimenta* (1516), as cited in Michael Bernhard, ed. and trans., *Johannes Turmair, genannt Aventinus, Musica rudimenta (Latein und Deutsch), das ist Anfangsgründe der Musik* (Facsimile Edition), part of *Veröffentlichungen der Gesellschaft für Bayerische Musikgeschichte* (Tutzing: Hans Schneider, 1980), preface. I wish to thank Dr. Michael Bernhard (Bayerische Akademie der Wissenschaften, Munich) for drawing my attention to this source. In past scholarship, this treatise has been incorrectly ascribed to Nicolaus Faber. (Cf. Ernst Praetorius, *Die Mensuraltheorie des Franchinus Gafurius und der folgenden Zeit bis zur Mitte des 16. Jahrhunderts,* vol. 2 of *Publikationen der Internationalen Musikgesellschaft* (Leipzig: Breitkopf und Härtel, 1905), p. 4. The question of authorship has also been discussed in Klaus Wolfgang Niemöller, "Ist Nicolaus Faber oder Johannes Aventin der Verfasser der 'Musicae rudimenta' (Augburg, 1516)?" *Die Musikforschung* 14 (1961): 184–185. The English translation is adopted from Young, "Franchinus Gaffurius," pp. 26–27.

67. Turmair, *Musicae rudimenta* (1516). Cf. Bernhard, *Johannes Turmair,* preface, and Young, "Franchinus Gaffurius," pp. 26–27.

68. It is in this context that Spataro's *Errori de Franchino Gafurio da Lodi da Maestro Joanne Spatario musico bolognese, in sua defensione, e del suo precettore Maestro Bartolomeo Ramis hispano subtilimente demonstrati* (Bologna, 1521), characterized by its prominent polemic undertone, needs to be judged. With regard to Gaffurio's *Theorica musice* (and the earlier *Theoricum opus musice discipline*), Spataro (justifiably) points out the following shortcomings: (a) the incor-

5:4 (rather than Boethius's 81:64), and the interval of the minor third by the ratio 6:5 (rather than Boethius's 32:27). This proposal in favor of the just intervals found the support of Spataro, who also advocated this division of the scale by just intervals. Gaffurio, a staunch supporter of Boethius and an ardent defender of tradition, did not accept the innovations of Ramos and Spataro but continued to subscribe to the Pythagorean tuning and also opted for a more conservative orientation in the grouping of the constituent pitches of the scale by hexachord.[69]

Translation and Annotations

This translation of Gaffurio's *Theorica musice* is based on the critical edition of the text[70] prepared from a copy of the 1492 printed edition at the Beinecke Rare Book and Manuscript Library of Yale University.

The principal aim of the translation has been to offer an idiomatic English rendition while remaining faithful to the language and style of the original. Special attention is given to the translation of technical terms, particularly to those which display multiple meanings.[71] In the *Theorica musice,* the translation of *vox, sonus, tonus,* and *corda* warrants special attention; *vox* is rendered as "voice" (*Theorica* 2.1.56) or "pitch" (*Theorica* 2.2.120); *sonus* as "sound"

---

rect application of the mathematical concept of incommensurability, as seen in the incorrect classification of the 5:3 ratio as an irrational proportion (*Theorica* 3.2.21, 31); (b) the inconsistent labeling of the 6:10 ratio as a superquadripartientsextal ratio (*Theoricum opus* 3.4), though without mentioning Gaffurio's emendation of this same ratio in the *Theorica* (3.3.68, 70, 79; 3.4.15–16) as a superbipartienttertial; (c) the confusion of the terms diesis and apotome (*Theorica* 4.3.64a, 73a); and (d) the overly dogmatic approach to the division of intervals on the monochord with regard to the placement of the smaller semitone below the larger semitone (*Theorica* 5.5). The Italian text of the *Errori* and an English translation are provided in Kreyszig, *Franchino Gaffurio's Theorica musice,* appendix 11.2.

69. For a more detailed discussion, see Heinrich Hüschen, "Kritik und Polemik in der Musiktheorie des 15. Jahrhunderts," *Festschrift Arno Forchert zum 65. Geburtstag am 29. Dezember 1985,* ed. Gerhard Allroggen and Detlef Altenburg (Kassel: Bärenreiter, 1986), pp. 41–47.

70. Kreyszig, "Franchino Gaffurio's *Theorica musice* (1492): Edition, Translation, and Study of Sources," Ph.D. diss., Yale University, 1989. Volume 3 contains the critical edition of the Latin text. My comparative critical edition of the *Theorica musice* (1492) and the *Theoricum opus musiche discipline* (1480) is in preparation and will be published by Hans Schneider of Tutzing, Germany in 1993.

71. Several authors have discussed these terminological issues of medieval music theory, thereby drawing special attention to changes in meaning of technical terms as they are manifest in different contexts or as they occur over a period of time. Cf. Margarete Appel, *Terminologie in den mittelalterlichen Musiktraktaten: Ein Beitrag zur musikalischen Elementarlehre des Mittelalters,* (Bottrop in Westfalen: Buch-und Kunstdruckerei Wilhelm Postberg, 1935); Hans-Heinrich Eggebrecht, *Handwörterbuch der musikalischen Terminologie* (Wiesbaden: Franz Steiner, 1972–); Hans Peter Gysin, *Studien zum Vokabular der Musiktheorie im Mittelalter: Eine linguistische Analyse* (Amsterdam: Frits Knuf, 1958).

(*Theorica* 2.1.6), "tone" (*Theorica* 5.1.4), "step" (*Theorica* 5.2.20), or "note" (*Theorica* 5.2.34); *tonus* as "tone" (*Theorica* 5.1.60) or "mode" (*Theorica* 5.8.41); and *corda* as "string" (*Theorica* 1.1.73) or "step" (*Theorica* 5.1.2). The original Latin of less common medieval terms appears in square brackets next to the translation. Medieval and Renaissance treatises on music generally include an abundance of particles, many of which frequently signal punctuation and paragraphing for the modern translator.[72] Thus the translator can simply ignore conjunctions like *et, nam,* and *enim* when they carry the meaning of a period or such particles as *scilicet* and *videlicet,* frequently represented by a colon in the translation.[73]

In the annotations to the translation, my primary concern has been to establish the sources to which Gaffurio was indebted for his theoretical thought, but I have also identified some parallel passages probably not known to him at the writing of the 1492 edition to give the reader further leads into the particular subject matter. Gaffurio's most likely source in these cases is given first, with the subsequent references arranged in order of proximity to Gaffurio's thought. Given the fact that Gaffurio had to rely largely on Latin works or on Latin translations of Greek treatises, the Greek sources are subordinated to the documents in Latin. In the *Theorica musice,* Gaffurio varied his mode of citing other authors' works, choosing among one of the following: paraphrase, gloss, summary, or verbatim quotation—with the latter reserved primarily for poetic texts. The task of identifying the sources is made difficult by the fact that Gaffurio was not consistent in naming the authors and treatises from which he derived his ideas—a failure endemic among humanist authors.[74] To make his prose uniformly clear and consistent, he blurred the boundaries of subtexts, putting another obstacle in the way of distinguishing borrowed material from interpretation and original exposition. The identification of these boundaries, which constitutes a vital part of the translator's task, is clarified in the translation through the consecutive numbering of sentences or principal clauses in bold in the text.

72. For a more detailed account on the Latin of Gaffurio, see: Walter Kurt Kreyszig, "Preparing Editions and Translations of Humanist Treatises on Music: Franchino Gaffurio's *Theorica musice* (1492)," in *Music Discourse from Classical to Early Modern Times: Editing and Translating Texts, (Papers Given at the Twenty-Sixth Annual Conference on Editorial Problems, University of Toronto, 19–20 October 1990),* ed. Maria Rika Maniates (New York: AMS Press, 1992), pp. 71–95.

73. This treatment of conjunctions and other particles is in agreement with the suggestions in Daniel Williman, "Translating Medieval Latin," *Translation Perspectives: Selected Papers, 1982–83,* ed. Marilyn Gaddis Rose as part of the publications of the National Resource Center for Translation and Interpretation (Binghamton, N.Y.: State University of New York at Binghamton, 1984), pp. 57–62. These guidelines are utilized in various modern translations, for example, in Bower, trans., *Boethius: Fundamentals.*

74. Miller, "Early Gaffuriana," p. 367.

ABBREVIATIONS

Abbreviations of classical sources are from Hammond and Scullard, eds., *Oxford Classical Dictionary* (2nd ed., 1984), and Günther Wille, *Musica Romana: Die Bedeutung der Musik im Leben der Römer* (Amsterdam, 1967).

Aaron	Aaron, Pietro
Lucid. in mus.	*Lucidano's in muscia*
Al. Mag.	Albertus Magnus
De quant.	*De quantitate*
Alc.	Alcaeus
Fragm.	*Fragment*
Alex.	Alexander [Polyhistor]
Not. on Phr.	*Notices on Phrygia*
Ambr.	Ambrose
Hex.	*Hexameron*
Anon.	Anonymous
Mus. ench.	*Musica enchiriadis*
Anon. XI	Anonymous XI
Tract de mus. plan.	*Tractatus de musica plana et mensurabili*
Ans.	Giorgio Anselmi
De mus.	*De musica*
Dia.	*Dialoghi*
Apul.	Apuleius
Flor.	*Florida*
Ar.	Aristophanes
Nub.	*Nubes (Clouds)*
Ran.	*Ranae (Frogs)*
Arch.	Archytas of Tarent
De math.	*De mathematica*
Arist.	Aristotle
De an.	*De anima*
De cael.	*De caelo*
Eth. Nic.	*Ethica Nicomachea*
Hist. An.	*Historia Animalium*
Metaph.	*Metaphysica*
Ph.	*Physica*
Poet.	*Poetica*
Pol.	*Politica*
Pr.	*Problemata*
Top.	*Topica*
Aristid. Quint.	Aristides Quintilianus
De mus.	*De musica*

Aristox.	Aristoxenus
Harm.	*Harmonica* (fragm.)
Ath.	Athenaeus
Deip.	*Deipnosophistai*
August.	Aurelius Augustine of Hippo
De civ. D.	*De civitate Dei*
De mus.	*De musica*
De ord.	*De ordine*
Bac.	Bacchius of Geron
Intro.	*Introductio artis musicae*
Barb., trans.	Ermolao Barbaro, translator
Them. Paraphr. (De an.)	*Themistius's Paraphrases on the De anima of Aristotle*
Them. Paraphr. (Metaph.)	*Themistius's Paraphrases on the Metaphysica of Aristotle*
Boet.	Anicius Manlius Severinus Boethius
De inst. arith.	*De institutione arithmetica*
De inst. mus.	*De institutione musica*
De uni. Trin.	*De unitate Trinitatis*
Boet., trans.	Boethius, translator
Arist. Cat.	*Aristotle, Categoriae*
Arist. Top.	*Aristotle, Topica*
Bryen.	Manuel Bryennius
Harm.	*Harmonica*
Bur., trans.	Giovanni Francesco Burana, translator
Bac. Intro.	*Bacchius of Geron, Introductio artis musicae*
Cap.	Martianus Capella
De nup.	*De nuptiis Philologiae et Mercurii*
Sat.	*Satyricon*
Cassiod.	Flavius Magnus Aurelius Cassiodorus Senator
Expo. in psalt.	*Expositio in psalterium*
De inst. div.	*De institutione divinarum litterarum*
De mus.	*De musica*
Inst. hum.	*Institutiones humanarum litterarum*
Var.	*Variae*
Cato	Marcus Porcius Cato
Agr.	*De agricultura*
Dist.	*Disticha*
Cens.	Censorinus
DN	*De die natali*

Frag.	*Fragmentum Censorini de musica*
Chalcid.	Chalcidius
in Tim.	*in Platonis Timaeum*
Cic.	Marcus Tullius Cicero
Her.	*Ad de ratione dicendi C. Herennium*
Leg.	*De legibus*
De Or.	*De oratore*
Rep.	*De republica*
Tusc.	*Tusculanea Disputationes*
Clem. Al.	Clemens Alexandrinus
Anecd. Oxon.	*Anecdotes Oxonensis*
Strom.	*Stromateis*
Cleon.	Cleonides
Intro. Harm.	*Introduction to Harmonics*
Com.	Peter Comestor
Hist.	*Historia Scholastica*
Con.	Conrad von Zabern
Nov. mus.	*Novellus musicae artis tractatus*
D'Abano	Pietro D'Abano
Concil.	*Conciliator differentiorum philosophorum et precipue medicorum*
Expo. Pr. Arist.	*Expositio Problematum Aristotelis*
Dam.	Damon
Areo.	*Areopagos* (fragm.)
Dem.	Demosthenes
Democr.	Democritus of Abdera
Guar.	*The Guarantees* (lost)
On Rhythm.	*On Rhythms and Harmony* (fragm.)
Diod.	Diodorus Siculus
Diog. Laert.	Diogenes Laertius
Vitae	*Vitae Philosophorum*
Dion. Hal.	Dionysius Halicarnassensis
Ant. Rom.	*Antiquitates Romanae*
Dios.	Dioscorides
Anth. Pal.	*Anthologia Palatina*
Epic.	Epicurus
Spic.	*Spicilegium*
Etym. Magn.	*Etymologicum Magnum*
Eur.	Euripides
Or.	*Orestes*
Fic., trans.	Marsilio Ficino, translator

Pl. Epin.	*Plato. Epinomis*
Pl. Leg.	*Plato. Leges*
Pl. Resp.	*Plato. Respublica*
Pl. Tim.	*Plato. Timaeus*
Finck	Hermann Finck
Pract. mus.	*Practica musica*
Gaf.	Franchino Gaffurio
Angelicum.	*Angelicum ac divinum opus musicae*
Extra.	*Extractus parvus musicae*
Flos mus.	*Flos musicae* (lost)
De harm.	*De harmonia musicorum instrumentorum opus*
Pract.	*Practica musicae*
Theorica	*Theorica musice*
Theoricum	*Theoricum opus musice discipline*
Gall.	Johannes Gallicus (Legrense)
Praef.	*Praefationcula in tam admirabilem quam tacitam et quietissimam novorum concinentiam*
Rit. can.	*Libelli musicalis de ritu canendi vetustissimo et novo*
Tacita	*Tacita nunc inchoatur stupendaque numerorum musica*
Gell.	Aulus Gellius
NA	*Noctes Atticae*
Glauc.	Glaucus of Rhegium
Hist. Chron.	*Historical Chronicles of Ancient Poets*
Guido	Guido of Arezzo
Epist.	*Epistola de ignotu cantu*
Microl.	*Micrologus*
Reg.	*Regula rhythmicae*
Handwörterbuch	*Handwörterbuch der musikalischen Terminologie,* ed. Hans-Heinrich Eggebrecht. Wiesbaden, Franz Steiner, 1972–
Hdt.	Herodotus
Nat. Hist.	*Natural History*
Heraclid.	Heraclides Ponticus
Coll.	*Collection of Facts Concerning Music*
Fragm.	*Fragment*
Hieron.	Hieronymous (Jerome)
Dial.	*Dialogus contra Pelagianos*
Hil.	Hilary of Poitiers
Tract.	*Tractatus super psalmos*

Hom.	Homer
Il.	*Iliad*
Od.	*Odyssey*
Hor.	Flaccus Quintus Horace
Ars P.	*Ars Poetica*
Sat.	*Satirae*
Hrab.	Hrabanus Maurus
De cler.	*De clericorum institutione*
De sac.	*De sacris ordine*
De univ.	*De universo*
Huc.	Hucbald
De harm.	*De harmonia*
De mus.	*De musica*
Isid.	Isidore of Seville
De eccl.	*De ecclesiasticis officiis*
Etym.	*Etymologiae sive origines*
Jac.	Jacques de Liège
Spec. mus.	*Speculum musicae*
Joseph.	Flavius Josephus
AJ	*Antiquitates Judaicae*
Lemprière	*Lemprière's Classical Dictionary* (London: Bracken Books, 1984)
Livy	Lucius Livius Andronicus
Op. om.	*Opera omnia*
Luc.	Lucian of Samosata
Dial. D.	*Dialogi Deorum*
Lucr.	Titus Carus Lucretius
De rer.	*De rerum naturae*
Op. om.	*Opera omnia*
Macr.	Ambrosius Aurelius Theodosius Macrobius
Somn. Scip.	*Commentarius in Ciceronis somnium Scipionis*
Mar. Vict.	Marius Victorinus (Aphthonius)
Ars gram.	*Ars grammatica*
March.	Marchetto of Padua
Lucid.	*Lucidarium in arte musice plane*
Mart.	Marcus Valerius Martial
Epigr.	*Epigrammes*
Spect.	*Spectacula*
MGG	*Die Musik in Geschichte und Gegenwart*, ed. Friedrich Blume. Kassel, 1949–73.
New Grove	*The New Grove Dictionary of Music and*

	Musicians, ed. Stanley Sadie. London: Macmillan, 1980.
Nicet.	Nicetus
De laude	*De laude et utilitate*
Nicom.	Nicomachus of Gerasa
Intro. to Arith.	*Introduction to Arithmetic*
Man.	*Manual of Harmonics*
Non.	Nonnus of Panopolis
Dion.	*Dionysiaca*
Ov.	Publius Naso Ovid
Fast.	*Fasti*
Met.	*Metamorphoses*
Pauly	*Pauly's Real Encyklopädie der Classischen Altertumswissenschaft*, ed. Georg Wissowa and Konrat Ziegler. 1893–1973; rpt. Stuttgart: Alfred Druckenmüller, 1962–1991.
Paus.	Pausanias of Lydia
Des. Graec.	*Descriptis Graeciae*
Pherec.	Pherecydes
Fragm.	*Fragment*
Phil.	Philolaus
De nat.	*De natura* (fragm.)
Philip.	Philippus
Anth. Pal.	*Anthologia Palatina*
Pl.	Plato
Epin.	*Epinomis*
Leg.	*Leges*
Phdr.	*Phaedrus*
Pol.	*Politica*
Resp.	*Respublica*
Soph.	*Sophistes*
Symp.	*Symposium*
Ti.	*Timaeus*
PL	Migne, *Patrologıae latina.* Turnhout, 1852.
Pliny	Gaius Plinius Secundus; the Elder
HN	*Historia Naturalis*
Pliny jun.	Gaius Plinius Caecilius Secundus; the Younger
Ep.	*Epistulae*
Plut.	Mestrius Plutarch
De an. proc. in Ti.	*De animae procreatione in Timaeo*

De mus.	*De musica*
De vi. Alc.	*De vita Alcibiadis*
Poll.	Pollux
Porph.	Porphyry
Harm.	*Harmonica*
Procl.	Proclus
Chrest.	*Chrestomathia*
Pros.	Prosdocimo de' Beldomandi
Par. tract.	*Parvus tractatulus de modo monacordum dividendi*
Tract. mus.	*Tractatus musice plane*
Ptol.	Claudius Ptolemy
Harm.	*Harmonica*
Quint.	Fabius Quintilian
Ist.	*Istitutio oratoria*
Strab.	Strabo
Geogr.	*Geographica*
Suda	Greek Lexicon, formerly known as *Suidas*
Suet.	Gaius Suetonius Tranquillus
Ner.	*Nero*
Theophr.	Theophrastus
De mus.	*De musica* (lost)
Tinct.	Johannes Tinctoris
De inv.	*De inventione et usu musicae*
Expo. man.	*Expositio manus*
Prop.	*Proportionale musices*
Trav., trans.	Ambrogio Traversari, translator
Diog. Laert. Vitae	*Diogenes Laertius, Vitae Philosophorum*
Ugol.	Ugolino of Orvieto
Decl.	*Declaratio musicae disciplinae*
Val. Max.	Valerius Maximus
Varro	Marcus Terentius Varro
Heb.	*Hebdomades vel de imaginibus* (lost)
De mus.	*De musica* (lost)
Rust.	*De re rustica*
Virg.	Publius Maro Virgil
Aen.	*Aeneid*
Ecl.	*Eclogues*
G.	*Georgics*
Vitr.	Marius Vitruvius Pollio
De arch.	*De architectura*

THE THEORY OF MUSIC

Franchino Gaffurio

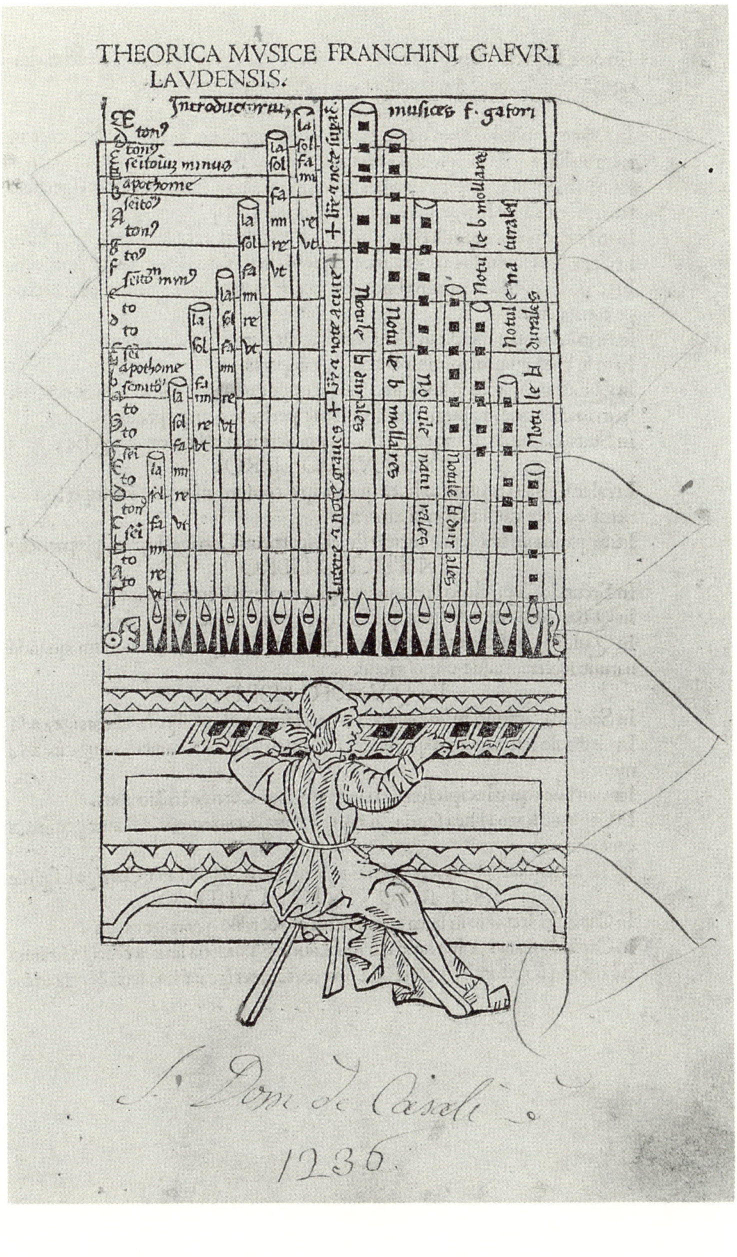

DEDICATION

1 To the magnanimous and most devout Lord Lodovico Maria Sforza, the Viscount, the duke of the most distinguished Principality of Bari and governor of the Milanese State: this theoretical work on the Art of Music of Franchino Gaffurio of Lodi, Music Director in the Choir of the greater church [cathedral].

2 To whose judgment ought I to assent, greatest Lodovico Sforza, ruler of your time? **3** To that of those who demand an account of leisure as well as of serious occupation and believe that a wasted day (when their work is dying of neglect) must be regained by twofold profit of diligence in making false charges? **4** Or to that of those who attribute the disgraceful mark of idleness and the ignominy of the base dullness of mind as a troublesome and peculiar burden to brute living beings? **5** This I do not yet know sufficiently well. **6** Those who propound what it fits a free man to do **7** and what everyone ought most greatly to avoid **8** put this before the eyes as if in a picture. **9** Because the study of the liberal arts and letters—among the first, the practice of writing—seem to suffice both notions [that is, what men should do and what they should not do], **10** I thought the account of my honorable leisure must be made valid. **11** I turned to free music by means of letters from its untidy neglect and abuse. **12** If I did not presume to outdo or equal the volumes researched by the ancients with more fruitful ingenuity, I have preferred to emulate in them (however little I have achieved of this) the most beautiful and not the most unpleasant. **13** I claim this work to be owed to my diligence and skill alone, that, if it may not be held in great estimation, **14** nevertheless it may be valued for its manifold and convenient utility. **15** For the ancients—by means of the agreeable tie to the other arts and by that faculty of association inborn to a noble sharpness of mind and by a luxuriant wealth of words as if by many stinging rays of light—blinded the eyes of [their] students with [their] lectures. **16** Almost all recent [authors] talk idly about this art out of either laziness or ignorance, with incomparable damage, and they now and then publish what is part inanity, part error, confounded and confused by a dry and barren style in a barbarous manner. **17** Abstaining from extremes, we proceed on a middle road through an easy path of explanation, that is, by writing in a pure and clear exposition with common, yet not poor, words, and we rejuvenate for a new beginning that which has become obsolete by age. **18** We have cleansed the obscure and false, we have added light to shadows, an entrance to subtleties, and faith to perplexities, **19** so that above all we desire that our vigilance be explored and inwardly examined, all the way to the finest details. **20** Moreover, we have improved upon the not unscrutinized little works for novices, **21** by revealing an easier way for those who long to learn, **22** not because our [approaches] are averse or inconsonant with theirs **23** but because it is not of little moment and makes a difference, in what way, how not discordantly and how clearly something is said. **24** In practice it happens that those who have cultivated the arts thoroughly, are ranked higher than the very inventors of them (because they give more results), **25** just as those who purify the gold drawn out from the interior of the earth **26** and distribute it for use among the people, surpass those who retrieve from the gold mines the more precious bronze. **27** Your Jacopo Antiquario, a consummate [man], has, however, reinforced my opinion (he "put the spur to a running horse" so to speak). **28** So great is both his authority and his piety, his erudition in all sciences, along with the highest purity of morals and of life, **29** that nothing is so open to question as how much he wants to research further. **30** Having been lured by his encouragement to writing, I most humbly dedicate this otherwise aimless work to you, gentlest of Princes, in good faith (the [faith], that is, which is owed entirely to a prince) and with my old devotion. **31** This is my own book, my little runaway, whose life is under the risk of a capital sentence. **32** It claims your [help] against its accusers, it names you its patron, and it recognizes you as an easy ruler. **33** Once you have published this noble compendium of doctrine from your rich resources, **34** and you promote the liberal arts and do not neglect the care of the artificers and of farmers, it prays for a humble place in your most venerable home. **35** You, a man most knowledgeable in the works of Mars, shut the creaking gates of the temple of Janus, and with a stillborn war have brought peace not only for Italy and Europe but also to the world. **36** In the firmest conviction of mankind, you have seized with both hands the palm of humanity, prudence, and generosity. **37** You think only wise thoughts. **38** You settle [all matters] with great deliberation. **39** You rule and govern with paternal and very great fortune. **40** Therefore the name "Moro" [mulberry tree] is bestowed upon you since it is the wisest of all trees. **41** Throughout the entire world, you recall the shunned virtues. **42** Now, you urge to be loved and to be sought out in the manner of the old emperors; **43** to your state you are a successful Augustus. **44** You challenge the learned and offer the very greatest rewards for labor. **45** To the eternal glory of your mighty name old heroes go with you on the heavenly path. **46** Even rude antiquity ascribed to them, who deserve the best from mankind, favors from the gods. **47** Therefore receive (as you are accustomed to do) with a cheerful countenance this

27 Secretary to the Governor of Milan. See also introduction, note 19.

musical volume, which has been written during many long nights with great labor by a man most dedicated to you. **48** May you cherish it in the company of your Antiquario, a man most erudite in literature, **49** and may the entire choir of singers applaud you with sweetest harmony. **50** Deign to inspect (as far as your occupations permit) what this latest inspiration has been capable of in me. **51** If (as I hope) [the work] will be well received, **52** I urge that, ornamented and perfected by your generous recommendation, **53** this book be regarded as worthy to reach the public on soaring wings. Farewell.

TO THE READER

Whoever you are whose fingers turn the leaves of these lessons as long as it be some good person who turns them—do I please you? Or is my work the only reward?

My intention was to help by pleasing—or is the weight of my merits empty?

I am unsure. Does gratitude despise the merit which I have gained?

There is no gratitude. Gratitude does not have any solid worth.

What does [Hrabanus] Maurus say? Will the Prince read this? He will read it. A deeper anxiety troubles [us].

That he reads it himself will be the fruit of praise.

Let him carefully read it through; someone will follow his example and read it also. My work is dedicated to this man; under his auspices may this work go forth.

All beginnings are hard. The subsequent joy usually softens spirits.

Thus the study which ensues aids by art.

Someone who occupies his leisure with mathematics passes time in such a way that there is no idleness.

Nevertheless, leisure is desirable to the provident.

Music ravishes the sky, earth, sea, bird, wild animals, stars, gods, humans and hell.

At first, music consisted by its nature in practice, but now it is examined in art.

Now finally it is a science. First was a desire [for music].

BOOK **ONE**

Chapter **One**
Musicians, [their] accomplishments, and also the excellence of musical knowledge

1 By reading and daily study I have perceived that in ancient times the discipline of music was embraced among the poets with the greatest honor of the most learned men—**1a** indeed, as Josephus **1b** and the Holy Scriptures relate that Jubal, from the tribe of Cain, was the first to have instituted [music] by the kithara and the organ, **1c** after he had searched out [this music] diligently from the various sounds of the hammers.

2 Many believe that others first distinguished themselves in various and diverse melodies and instruments. **3** Heraclides Ponticus, **3a** whom Plato, before sailing to Sicily, had appointed head of the Academy, reported **3b** that Amphion, the son of Jupiter and Antiope, was the original creator of the playing of the kithara and of the song with kithara [accompaniment], and he was instructed on the kithara by his father, Jupiter. **4** Others believed that the eminent Linus first discovered the lyre; **4a** he was from Calchos [on the island] of Euboea, the son of Apollo and Terpsichore, or, according to others, born from Amphimarus or Mercury as father and from Urania as mother. **5** So Linus wrote lamentations in lyrical song, as Anthes from Anthedonia, who at the same time also composed hymns. **6** Furthermore, many other people are said to have lived during the same time, as well as Pierius from the Macedonian mountain of Pierius, who composed poems on the Muses. **7** We read that Philamon of Delphi, who first established the times of that temple when it was at Delphi, proclaimed the origin of Leto, Diana, and Apollo in melody. **8** Thamyras, too, the son of Philamon himself, is said to have sung more sonorously and

1a Joseph. *AJ* 1.64, 68–72.
1b Ugol. *Decl.* 5.31.2, 4; Genesis 4.20–21.
1c Ugol. *Decl.* 5.31.3.
3, 3b Plut. *De mus.* 1131F–1132A; Heraclid. *Fragm.* 157 (ed. Wehrli).
4, 5–6 Plut. *De mus.* 1132A. In the transmission of this legend, Gaf. perhaps followed Cens. *DN* 12. Concerning Linus, see Hannelore Thiemer, *Der Einfluss der Phryger auf die altgriechische Musik,* vol. 29 of *Orpheus-Schriftenreihe zu Grundfragen der Musik,* ed. Martin Vogel (Bonn-Bad Godesberg: Verlag für Systematische Musikwissenschaft, 1979), pp. 91–93.
7 Plut. *De mus.* 1132A; Pherec. *Fragm.* 120 (ed. Jacoby).
8 Plut. *De mus.* 1132A; Hom. *Il.* 2.594–600.

melodiously than any in that day and age. **9** Demodocus of Corfu, a not at all unknown musician among the ancient writers, **9a** as Heraclides himself mentions, **9b** wrote songs about that overthrow [of Troy] **9c** and about the adultery of Venus and Mars. **10** Phemius [from the island] of Ithaca described in a poem the return of those who came back from Troy with Agamemnon.

11 Truly, they all wrote in verses, to which they later added melodies, as Terpsichorus and the remaining ancient lyric poets and poets of songs were accustomed to do. **12** Terpander, the first to write nomoi for the lyre, built the heptachord into the lyre; **12a** nevertheless several writers credited Philamon with adding separate melodies to fit each of his nomoi of kitharody **12b** and also to fit the verses of Homer. **13** In the same manner Clonas, the elegiac poet and the first epic [poet], established the aulodic nomoi and harmonies. **14** Following him, Polymnestus of Colophon pursued the same compositions in verse.

15 But two nomoi for kithara were discovered much earlier than those for the aulos and were established during the times of Terpander. **16** For Terpander himself was the first who named the nomoi and melodies for the kithara, now Boeotian, now Aeolian, another time Trochaic, and then acute, yet another time Cepionic and Terpandric. **17** [Terpander] also wrote prooemia [preludes] for the kithara, **17a** although Alexander—**17b** in his collec-

9, 9b Plut. *De mus.* 1132B; Hom. *Od.* 8.499–520; 8.266–366.

9c Gaf. changed the original Plut. text which mentions the marriage of Aphrodite (latinized Venus) and Hephaestus (latinized Vulcanus).

10 Plut. *De mus.* 1132B; Hom. *Od.* 1.325–327.

12 Though not explicitly expressed here, this statement is in clear reference to a later passage in Plut. *De mus* 1132D; cf. Poll. 4.65.

12a Plut. *De mus.* 1132C in a slightly altered form ("Terpander also, who was a composer of nomoi sung to the cithara, set to music in each nome hexameters"); see Procl. *Chrest.* 45.

12b Plut. *De mus.* 1132C.

13–14 Plut. *De mus.* 1133A.

15–17 Plut. *De mus.* 1132D; Poll. 4.65.

16 The Latin text has "Boetium," obviously a misspelling in the *Theorica musice,* which should read "Boeotium" in reference to that nomos named after the district of Boeotia in central Greece.

17a Plut. *De mus.* 1132F; Alex. *Not. on Phr.* (*Fragm.* 77, ed. Jacoby).

17b Humanist translators of scientific texts tended to follow the Scholastic process of word-for-word conversion of Greek into Latin, maintaining as much of the word order, inflection, and syntax as possible. For further information on this point, see Deno John Geanakoplos, *Constantinople and the West* (Madison: University of Wisconsin Press, 1989), pp. 82, 283. For specific examples, see Palisca, *Humanism*, p. 120, and Thomas J. Mathiesen's response to Claude V. Palisca, "The Impact of the Revival of Ancient Learning on Music Theory" in *International Musicological Society: Report of the Twelfth Congress Berkeley 1977,* ed. Daniel Heartz and Bonnie Wade (Kassel: Bärenreiter, 1981), p. 880.

The Greek refers to Alex.'s *Not. on Phr.* (ἐν τῇ Συναγωγῇ τῶν περὶ Φρυγὶας [literally in the collection of things concerning Phrygia]) wherein it is stated that (1) Olympos brought instrumental music (κροὺματα) to the Greeks, (2) so did the Idaean Dactyls, (3) Hyagnis was

tion of those who were versed on thc Phrygian rhythms—**17c** narrates that the Phrygian Olympus was the first who carried forth the art of aulos playing and dactylic verse composing to the Greeks—**17d** and that, indeed, during the times of the reign of the Phrygian Midas it was he who first invented the oblique aulos.

18 But the one who first employed the aulos [a1v] was Hyagnis, and **18a** thereafter his son Marsyas, who, carried away by pride, boasted that he was the inventor of the art of aulos playing, **18b** to the extent that he challenged Apollo to a musical contest; **19** Apuleius and Diodorus related that Marsyas was conquered by Apollo and skinned alive. **20** In fact, afterward Apuleius and Diodorus mention that there was an Olympus, a disciple of this Marsyas, who was born before the Trojan war **20a** and therefore older than the Phrygian Olympus, whom we have mentioned. **21** Indeed, Apuleius and Diodorus say that Terpander imitated the verses of Homer but the melodies of Orpheus. **22** Nevertheless Orpheus is said to have imitated no one, inasmuch as there was not yet anyone besides the aulodic poets. **23** But the songs Orpheus sang have no resemblance to the songs of those poets. **24** On the other hand, it is well known that not long after Terpander there was Clonas, a composer of nomoi for the aulos—according to Archades a citizen of Tegea, but according to the Boeotians a Theban. **25** After these two, Terpander and Clonas, Archilocus is placed, **25a** although certain historians report that it was a certain Ardalus of Troezen prior to Clonas, who established aulodic music. **26** However, others claim that it was Polymnestus.

---

the first to play the aulos, and so on. The interlocking indirect discourse has caused the Latin translator to misconstrue the phrase as *ἐν τῇ Συναγωγῇ τῶν περὶ Φρυγὶας κροὺματα* . . . instead of *ἐν τῇ Συναγωγῇ τῶν περὶ Φρυγὶας* kroùmata . . . ; he consequently takes it as "in (*ἐν*) eorum (*τῶν*) collectione (*τῇ Συναγωγῇ*) Phrygii (*περὶ*) pulsationes (*κροὺματα*) versati sunt" (because the final coordinating verb seems to be missing in the Greek, the translator supplies it from the context). This information was kindly communicated to the translator by Professor Thomas Mathiesen on 4 February 1992.

17c Plut. *De mus*. 1132F; Clem. Al. *Strom*. 1.16.76.6 (ed. Stählin); Clem. Al. *Anecd. Oxon*. (ed. Cramer, Vol. 4, p. 400.19); *Suda* (ed. Adler, vol. 3, p. 522.22).

18 Plut. *De mus*. 1132F; Dios. *Anth. Pal*. 9.340; Non. *Dion*. 41.374.

18a Plut. *De mus*. 1132F; Philip. *Anth. Pal*. 9.266; Apul. *Flor*. 3; Non. *Dion*. 10.233.

18b Apul. *Flor*. 3; Luc. *Dial. D*. 16.2 (cf. Thiemer, *Der Einfluss der Phryger,* pp. 53–54).

19 Apul. *Flor*. 3; Diod. 3.58 (cf. Thiemer, *Der Einfluss der Phryger,* p. 54).

20 Plut. *De mus*. 1132F; Alex. *Not. on Phr*. (*Fragm*. 77, ed. Jacoby).

20a Cf. above 1.1.17a.

21 Plut. *De mus*. 1132F; Alex. *Not. on Phr*. (*Fragm*. 77, ed. Jacoby).

22 Plut. *De mus*. 1132F; *Orphicorum Fragm*. 56–58 (ed. Kern).

23 The assumed reference is to Glauc. (*Hist. Chron*.), who seems to refer to Ardalus of Troezen (cf. Plut. *De mus*. 1133A and below to 1.1.25).

24–25 Plut. *De mus*. 1133A.

25a Plut. *De mus*. 1133A; Pliny *HN* 7.204; Paus. *Des. Graec*. 2.31.3.

26 Plut. *De mus*. 1133A.

27 The kitharody of Terpander was preserved entirely in simple form all the way to the generation of Phrynis of Mytilene. **28** At that time this very experienced Phrynis left behind this kind of kitharody handed down from Terpander, under the guidance of kitharist Aristoclides. **29** This Phrynis was the first to begin to alter this ancient and simple music and to transpose [it] into new harmonies and rhythms. **30** Truly, Olympus the Phrygian, **30a** whom we have just mentioned previously, **30b** laid down the aulodic nomos in honor of Apollo. **31** This Olympus was one of those who proceeded from that first Mysian Olympus, the disciple of Marsyas, by whom the nomoi and melodies to the gods were instituted, which afterward he carried to Greece. **32** Yet others contend that the Polycephalus [nomos] was not the work of the Phrygian Olympus but of Crates, whom Olympus taught. **33** The Italian Glaucus [of Rhegium], however, in his *Historical Chronicles of the Ancient Poets* asserts that the Chariot nome **33a** was a melody of Olympus, which also may be understood from Terpsichorus Hymenaeus, **33b** for Glaucus imitated neither Orpheus nor Terpander nor Archilochus nor Thales, but solely Olympus. **34** Likewise, some say that this nomos was discovered by the Mysians, who were among the ancient auletes. **35** For they sang elegies made up of melodies. **36** They also handed down that there was Sacadas, an Argive, a composer of these kinds of melodies, a good composer and a three-time winner at the Pythian games.

37 There are those who seek this invention farther back. **38** They would have it that the inventor was Apollo, who first invented not only the kithara but also the aulos long before Marsyas or Olympus or Hyagnis. **39** At least, this fact is evident from these choruses and the sacrifices, which, as Alcaeus writes in a certain hymn, antiquity offered to the god Apollo with auloi. **40** In addition, his statue, which was at Delos, is said to have held a bow in its right hand, and in the left the Graces, who each bore in their hands separate instruments of music, one the lyre, the other auloi. **41** The Grace in the middle held a syrinx to her mouth.

27 Plut. *De mus.* 1133A; Pol. 4.66; Procl. *Chrest.* 46.
30, 30b, 31 Plut. *De mus.* 1133D; see also 1132F.
30a Cf. above 1.1.17a.
33, 33b The Latin text has "harinatium," an incorrect Latin transliteration of the Greek (latinized "harmatium"). Plut. *De mus.* 1133F. Concerning the explanation of this nomos, see *Etym. Magn.* 145.25–47 and Eur. *Or.* 1384.
34 Plut. *De mus.* 1133F.
36 Plut. *De mus.* 1134A; Paus. *Des. Graec.* 10.7.4.
38 Plut. *De mus.* 1135F; cf. also 1133F.
39 Plut. *De mus.* 1135F–1136A; Alc. *Fragm.* 3 (Bergk, *Poet. Lyr. Gr.* 3, p. 147) and *Fragm.* 307 (Lobel and Page, *Poet. Lesb. Frag.*, p. 259).
40–41 Plut. *De mus.* 1136A; Paus. *Des. Graec.* 9.35.3.

42 The Egyptians say that Mercury was the first observer of the course of the stars and that he invented the harmonies to correspond. **43** He built a three-string lyre [as] an image of the three seasons of the year, **44** that is, he instituted three pitches, the high, the low, and the middle, deriving the high from the summer season, the low from the winter, and the middle from the spring. **45** The Egyptians also report that Osiris [a2r] was accustomed to be delighted by musicians and choirs. **46** A multitude of musicians and nine adolescent virgins, accustomed to sing and to be taught other things, used to follow Osiris. **47** These virgins the Greeks called Muses. **48** Apollo is said to have taught these virgins, for which reason he is called "musicus."

49 The Arcadians were said to have been the first to bring these instruments, which were called the lyre, the trigonon, and the lydi, into Italy. **50** In fact, this people formerly made use of no other skill beyond the pastoral pipes. **51** They also relate that Clonas was the composer of a tripartite nomos which is made up of the Dorian, the Phrygian, and the Lydian modes. **52** Finally, it is recorded that Thales and Xenodamus, Xenocritus and Polymnestus, **52a** and also Zetus, **52b** Arion, Timotheus, Antigenidas, Polixenus, and many others were famous during this period of time.

42–44 Cassiod. *Var.* 2.40.14. "Egyptii" is perhaps a reference to the Alexandrian astronomer Hypsicles.
51 Plut. *De mus.* 1134A.
52 Plut. *De mus.* 1134B–C. Plut. provides the native town of these authors, namely, "Thaletus of Gortyn, Xenodamus of Cythera, Xenocritus of Locri, Polymnestus of Colophon, and Sacadas of Argos." Thaletas of Crete is mentioned in Boet. *De inst. mus.* 1.1.181.23 (cf. Wille, *Musica Romana,* p. 659), in Varro, *De mus.* (lost) (cf. Wille, *Musica Romana,* p. 446), and in Cap. *De nup.* 9.493.6–7 (cf. Wille, *Musica Romana,* p. 446). The remark of the latter author is cited in *Theorica* 1.1.86. Together with Xenocritus, Xenodamus, Polymnestus, and Sacadas (all mentioned in Plut. *De mus.* 1134B–C and in *Theoricum* 1.1.13–14; the name Sacadas is deleted in the *Theorica*), Thaletas belonged to the second School of Music in Sparta, where he was credited with instituting the *gymnopaediai,* an annual festival held at Sparta to honor Apollo with gymnastic exercises and dances (cf. Solon Michaelides, *The Music of Ancient Greece: An Encyclopaedia* [London: Faber & Faber, 1978], pp. 125, 330). Polymnestus of Colophon is also mentioned in *Theorica* 1.1.14, adopting a statement from Plut. De mus. 1132C, 1133A, as well as in *Theorica* 1.1.3, adopting a statement from Plut. De mus. 1131F–1132A. Both Pindar (frag. 178) and his contemporary, the comic dramatist Cratinus (frag. 305), mention Polymnestus (cf. Warren Anderson, "Polymnestus of Colophon," *New Grove* 15:54).
52a Cic. in his *Her.* 2.27.43, mentions Amphion together with his twin brother Zetus as son of Jupiter and Antiope (cf. Wille, *Musica Romana,* p. 552).
52b Pliny (*HN* 9.8.28) and Hdt. (*Nat. Hist.* 1.23) mention the famous account of Arion, the kitharode. Together with Terpander, Arion is cited by Boet. (*De inst. mus.* 1.1.185.17–20) in reference to the healing of grave diseases through song. Gaf. gives the respective accounts in his *Theorica* 1.1.65, 90. Gell. (*NA* 16.19.2–17) and Ov. (*Fast.* 2.83–118) also include references to the Arion mythos (cf. Wille, *Musica Romana,* pp. 553–554). Cap. (*De nupt.* 9.480.16) mentions Arion, Orpheus, and Amphion in a blending of their lyres (cf. Wille, *Musica Romana,* p. 640). Both Procl. (A.D. 412–485) (*Chrest.* 12) and Arist. (*Poet.* 1449a 1.11)

53 These [musicians], the more they differ among themselves, the more they impart a greater inference of antiquity and dignity to music. **54** On account of the very remote antiquity, the author of a human invention ought to be all the more uncertain. **55** However the dignity of the thing invented draws many people to love it, so that individuals wish to be called the inventor, if they can—something that happens not only among musicians but also in philosophy and the best other disciplines. **56** But how great is the force and power of harmonic proportion, which, **56a** as Pliny says in the last [chapter] of the second [book], **56b** forces the natures of things to agree with themselves, ought to be succinctly discussed.

57 It has been fancied indeed, that Orpheus and Amphion drew together animals who lack reason and stones and forests by songs. **58** But [the theorists] hand down that after the Muses had been fitly brought together and their song shown to the world, many men were so taken by this delight that they neglected food and drink while singing, and so they imprudently perished, from whom afterward the race of the tree-crickets was born. **59** This [animal] received this gift from the Muses, so that it would not require nourishment but could sing without food or drink as long as it lived.

60 We read that there is no heart so harsh and so bitter that it is not excited by the feeling of such pleasures. **61** Perhaps the first musicians attracted barbarous races that lacked the cultivation of reason or, like a rock, were insensitive to feeling to the sense of pleasure by singing; **61a** Quintilian teaches this in his praise of music **61b** and Horace expresses it in the *Art of Poetry* in these verses:

---

mention Arion's association with the dithyramb (cf. Warren Anderson, "Arion," *New Grove* 1:581). Timotheus's innovations and controversial disputes, the latter largely resulting from the former, are mentioned by Pherecrates in his comedy *Cheiron* (cf. Michaelides, *The Music of Ancient Greece,* p. 334), as cited in Plut. *De mus.* 1141C ff. and in Boet. *De inst. mus.* 1.1.182.1; 1.1.183.11 (cf. Wille, *Musica Romana,* p. 659) and *De inst. mus.* 1.20.208.29 (cf. Wille, *Musica Romana,* p. 671). Arist. in his *Metaph.* 1.993B praises Timotheus's accomplishments (cf. Michaelides, *The Music of Ancient Greece,* p. 334). Gaf.'s more specific remark to Timotheus (cf. *Theorica* 1.1.79) appears to be based on the previously mentioned citation from Plut.'s *De mus.* Gaf.'s mention of the name Antigenidas presumably stems from Plut.'s more elaborate comments on this aulete's significance and reputation (cf. Plut. *De mus.* 1138B). The sophist Polixenus is mentioned by Diog. Laert. in his *Vitae* 2.76 and by Pl. in his *Epin.* 2.314 cd (cf. Willy Stegemann, "Polyxenos," in: *Pauly* 21,2, pp. 1857–1859. "Many others" (*Theorica* 1.1.52b) is a reference to Sacadas and Pratinas (cf. Plut. *De mus.* 1134A. Gaf. mentions Sacadas above (cf. 1.1.36).

56, 56b Pliny *HN* 2.113.

58 Pl. *Phdr.* 258e6–259d8. For an explanation of this passage, see G. R. F. Ferrari, *Listening to the Cicadas: A Study of Plato's Phaedrus, Cambridge Classical Studies,* ed. M. F. Burnyeat et al., (Cambridge: Cambridge University Press, 1987), p. 26. A further reference to the tree-cricket (*cicada*) is provided below in 2.1.58.

61a Quint. *Ist.* 1.10.

> **62** And he, the venerable interpreter of the gods, hindered those men who reside in forests from slaughter and wild habits.

63 Orpheus is said in this connection to calm tigers and swift lions, **64** and Amphion, the founder of the city of Thebes, to have moved stones by the sound of the lyre and by a persuasive request to go where he wished. **65** A dolphin, also, persuaded of the man's friendship to him by music of his kithara, carried the kitharode Arion, who had been thrown into the ocean, safely to the coast, **65a** as Herodotus relates in the first [book] of his *History,* **65b** and also Pliny in the ninth [book] of his *Natural History*. **66** They mention that in Lydia there are islands of nymphs, as also the more recent of those who asserted this fact, Varro, is testified to have himself observed. **67** These nymphs advance from the land to the middle of the lake by the sound of the tibia, at first moving in a circle and from this place retreating to the shore. **68** They say that the serpents are attracted by song and burst asunder, **68a** and the Indian elephants are detained, delighted by the instrumental music, and also the deer are captured by means of pipes. **69** Martianus Capella wrote that the swans are attracted by the music of the kithara. **70** Moreover, many birds, such as goldfinches, nightingales, and others, practice song as if by a certain artistic discipline, **71** and almost all [a2v] birds and likewise both terrestrial and aquatic beasts run down into a net freely, allured by song, as the authority of a wise man discloses in this poem. **72** "The flute sounds pleasantly while the fowler snares the bird."

73 The very powerful tongue of [Virgil] Maro celebrates Musaeus, the son of Orpheus and Nature, most famous in the art, saying that he has been placed among the dead with the highest happiness, because he animated the happy souls in the Elysian fields by the striking of seven strings. **74** [Virgil] indicates that he enjoys the highest reward to whom it is given to banquet on the

62–64 Hor. *Ars P.* 391–396.
65 Concerning Arion, see C. M. Bowra, "Arion and the Dolphin," *Museum Helveticum: Schweizerische Zeitschrift für klassische Altertumswissenschaft 20* (1963):121–134.
65a Hdt. *Nat. Hist.* 1.23.
65b Pliny *HN* 9.8.28; Varro, *De mus.* (lost) (cf. Wille, *Musica Romana,* p. 415).
66–67 Cap. *De nupt.* 9.493.20–494.4; Pliny *HN* 2.95.209; Varro *Rust.* 3.17.4. The original account comes from Theophrastus's collection of miraculous stories (cf. Wille, *Musica Romana,* p. 416).
68 Cap. *De nup.* 9.493.17.
68a Cap. *De nup.* 9.493.13–15. For a further reference to elephants see above 1.6.19.
69 Cap. *De nup.* 9.493.12–13; Macrob. *Somn. Scip.* 2.3.10; Cassiod. *De mus.* 1212A; Isid. *Etym.* 3.17.3; Varro, *De mus.* (lost) (cf. Wille, *Musica Romana,* p. 415).
72 Cato *Dist.* 1.27.2. I would like to thank Dr. Leofranc A. Holford-Strevens (Oxford) for his identification of this verse in a letter of 1 August 1992.
73–74 Cassiod. *Var.* 2.40.60–65.

sweetnesses of this discipline. **75** Surely, one must admire and also believe that David's lyre drove out the devil, **75a** and that he held control over the spirits through sound; **76** and after the kithara sounded three times, the king, whom the inner foe tortured in a terrible manner, returned to a state of inner freedom. **77** Indeed, the song of no musical instrument is more powerful in moving minds than is the pleasant resonance of a concave kithara. **78** In his *Letter to Boethius,* Cassiodorus judges that a string is so called because it easily moves where a so great concord of pitches is made out of diversity, so that a neighboring string which has been struck causes another [string] to tremble spontaneously, although it has not chanced to come in contact with it.

79 Timotheus, who distinguished himself by far over all musicians, as often as he wished excited and inflamed the souls of men by a vehement harmony and, conversely, calmed them with a soft and gentle harmony. **80** During a banquet given by Alexander, [Timotheus] played the Phrygian mode and is said to have repeatedly excited the king to such a degree that he hastened to seize arms and then, with a change in melody, withdrew back a second time to the festivities and banquet. **81** The Pythagoreans calmed with melodies the soul of a youth who was furious to burn the house of a rival in which [his] mistress was detained. **82** Damon tamed the intoxicated and in addition quite violent and freakish youths by gravity of melody. **83** Surely it was he who, having ordered the aulete to play the spondaic [melody], broke the insanity of a confusion caused by intoxication. **84** When the great mass of the common people raged confusedly against the ruling fathers of the cities, music frequently inhibited those inflamed by sedition. **85** Xenocrates freed frantic people by instrumental music.

75 Cassiod. *Var.* 2.40.94; *Samuel* 1. 16.23; 19.9.
75a Cassiod. *Var.* 2.40.94.
76 Cassiod. *Var.* 2.40.94–103. Cassiod. presumably derived this information from Nicet. *De laude* 3.
77–78 Cassiod. *Var.* 2.40.94–103.
78 The wordplay here, based on the parallel between *corda* (string) and *concordia* (concord), is intended to emphasize the etymological similarity between "string" and "concord." Thus, a *corda* is so called because it moves in sympathy to a *concordia* of pitches. This information was kindly communicated to the translator by Professor Thomas Mathiesen on 4 February 1992.
79 Though not a direct quotation from the literature, Gaf.'s reference to Timotheus of Milet seems to summarize Plut.'s remarks in his *De mus.* 1141C–F.
80 Tinct. *De invent.* 3 (Weinmann, p. 39); see also Gaff. *De harm.* 4.5. On the widespread dissemination of the Timotheus legend, see Nan Carpenter, "Spenser and Timotheus: A Musical Gloss on E. K.'s Gloss," *Publications of the Modern Language Association* 7 (1956): 1141–1151.
81 Boet. *De inst. mus.* 1.1.185.3–4.
82–83 Cap. *De nup.* 9.492.15–19.
84 Cap. *De nup.* 9.492.13–15
85 Cap. *De nup.* 9.493.5.

86 It is well known that Thales of Crete drove away illnesses and contagious diseases by the sweetness of the kithara. **87** Hierophilus examined the veins and pulses of the ill by comparing rhythms [heartbeats]. **88** The physician Asclepiades often restored the senses of the frenetic, disturbed by illness, to health through harmony. **89** The ancients cured fever and wounds with song. **90** Terpander and Arion delivered the peoples of Lesbos and Ionia from very grave diseases by the assistance of song. **91** It was believed that when there was greatest suffering under the hips, then the pains were diminished if the aulete sounded light melodies upon them. **92** And Theophrastus was of the opinion that the aulete, by playing skillfully and melodiously, heals the bites of vipers. **93** Moreover, Democritus, in his book *On Epidemics*, teaches that the playing of the auloi was medicine for the many diseases of men. **94** Truly, so great is the affinity between the minds and the bodies of men and therefore also between the ailments of mind and body and [their] remedies.

95 However, the cure of music did not fail for rough and warlike men and great leaders. **96** Epaminondas is said to be the first in Greece in singing to the lyre. **97** Moreover, they relate that Claudius Nero was so delighted by music that he wore a sheet of lead on his breast for the sake of his voice, in order to cure and preserve it. **98** Nero also summoned the kitharist Terpnus, famous at that time above all others, to grant him a great reward.

99 Let us no longer demand examples! **100** We see that the first men and

86 Cap. *De nup.* 9.493.6–7; Tinct. *De invent.* 1.5.195–196.

87 Cap. *De nup.* 9.493.7–8; Pliny *HN* 29.1.6; Cens. *DN* 12.4; Tinct. *De invent.* 1.5.196–197; Varro, *De mus.* (lost) (cf. Wille, *Musica Romana,* p. 415).

88 Cens. *DN* 12.4; Cap. *De nup.* 9.492.11–13; Macrob. *Somn. Scip.* 2.3.9; Boet. *De inst. mus.* 1.1.184.7; Cassiod. *De mus.* 9.1212B; Isid. *Etym.* 4.13.3 (cf. Wille, *Musica Romana,* p. 711); Varro, *De mus.* (lost) (cf. Wille, *Musica Romana,* p. 415); Cap. *Sat.* 9; Guido *Microl.* 14.

89 Cap. *De nup.* 9.493.1–2. A similar practice is transmitted in connection with Hom.'s healing of bleeding wounds through magic poetry (cf. Pliny *HN* 28.2.21, as cited in Wille, *Musica Romana,* p. 446).

90 Boet. *De inst. mus.* 1.1.185.17–20 (cf. Wille, *Musica Romana,* p. 446). This reference was not included in Varro's *De mus.* (lost) (cf. Wille, *Musica Romana,* p. 659; also Walther Miekley, *De Boethii libri de musica primi fontibus,* Jena, 1898, p. 29).

91 Gell. *NA* 4.13.1; Pliny *HN* 28.2.21; Varro *De mus.* (lost) (cf. Wille, *Musica Romana,* pp. 415, 446, 659; Miekley, *De Boethii libri de musica,* p. 27). Boet., in his *De inst. mus.* 1.1.185.20 provides a more specific reference to the aulos player Ismenias (cf. Wille, *Musica Romana,* p. 446).

92 Gell. *NA* 4.13.2.

93 Gell. *NA* 4.13.3 Of Democritus's numerous treatises, some of which are noted for their spurious attribution, none are extant (cf. *Pauly* 5.1, p. 137). Perhaps the reference here is indeed to a two-volume set of Hippocrates entitled *On epidemic diseases* (cf. Th. Meyer-Steineg and Karl Sudhoff, *Geschichte der Medizin im Überblick mit Abbildungen,* 2nd ed. (Jena:Gustav Fischer, 1922), p. 54).

94 Gell. *NA* 4.13.4.

97 Pliny *HN* 3.4.18.166 (cf. Wille, *Musica Romana,* p. 480).

98 Suet. *Ner.* 20.1 (cf. Wille, *Musica Romana,* p. 339).

leaders in this our time surpass in musical sweetness those ancients who are much more subdued. **101** Indeed, there are those who, for the practice of the art, wish musicians to assist with playing to ease anxiety at banquets and games and to follow along also in almost all labors.

102 The army of the Lacedaemonians was inflamed by the auloi and the musical modes. **103** The Cretans fought to [the sound of] kitharas. **104** The Amazons practiced arms to the reed pipes. **105** One [of the Amazons], who had come desirous of meeting Alexander, when she greeted him was honored with a piper, so that she could leave graced with a great gift. **106** The Lacedaemonians in Greece and in Italy the Sabaristae sent auletes before them into battle.

107 What else do drums and trumpets accomplish in our legions and in the military camps? **108** The more vehement their sounds blending harmoniously together, the stronger the willpower of the warriors in combat. **109** From this I determine that Roman glory, above all others, stood out in war, **109a** and this is not surprising since even to timorous animals the clamor of victory is musical, **109b** as Virgil pointed out in the *Georgics* in that passage about bees:

> **110** If, however, [the bees] have gone forth to the battle—for often a dissension between the two kings occurred with great commotion—straightaway they learned the minds of the crowd and the hearts that trembled at war, perhaps at a distance, for a martial noise of raucous bronze reproves the hesitant ones. **111** And a voice is heard which imitates the blaring sounds of trumpets. **112** Then [the bees] assemble in a state of alarm among themselves and flash with their wings. And they sharpen their stings on the beaks and flex their muscles. **113** And they will swarm around the king and his guard, **114** and summon the enemy with great clamor.

115 This musical science has been placed among all holy festival celebrations with the utmost respect for the sake of pleasing the gods. **116** For the Marmedians and the Aones play upon reed instruments in praise of the heavenly gods. **117** The Egyptians offered the pandoura along with pastoral pipes to their gods. **118** The Cretans and Lacedaemonians showed themselves

103 Cap. *De nup.* 9.491.18; Quint. *Ist.* 1.10.14 (cf. Wille, *Musica Romana,* p. 414); Gell. *NA* 1.11.6, based on Varro, *De mus.* (lost) (cf. Wille, *Musica Romana,* p. 578).
104 Cap. *De nup.* 9.492.1–5; Pliny *HN* 8.42.157 (cf. Wille, *Musica Romana,* p. 414).
105–106 Cap. *De nup.* 9.492.1–5; Varro *De mus.* (lost) (cf. Wille, *Musica Romana,* p. 414).
107 Cap. *De nup.* 9.492.5–7; Boet. *De inst. mus.* 1.1.186.28–30; Quint. *Ist.* 1.10.14; Cens. *DN* 12.3; Macrob. *Somn. Scip.* 2.3.9; Varro *De mus.* (lost) (cf. Wille, *Musica Romana,* p. 415).
110–114 Virg. *G.* 4.67–76.
116–117 Cap. *De nup.* 9.491.8–11.
118 Cap. *De nup.* 9.491.18–20.

humble petitioners to the Muses with the kithara and the auloi before they approached the fates of wars. **119** King Nebuchadnezzar ordered all types of musicians for the statue which he erected. **120** The tenth book of Strabo indicates that Plato, in the seventh [book] of [his] *Laws,* handed down that hymns and praises of the gods were sung, blended with prayers, and that dances and other kinds of song were dedicated as holy to the gods, and that every type of music was thought the work of the gods. **121** Beyond this, these sacred rites were once sent from the Hyperboreans to Delos with auloi and syrinx and kitharas—as was the custom for this to be done.

122 When the tibia players, who, among the Romans, had been traditional since ancient times in the temple of Jupiter, were prohibited by the subsequent censors, they indignantly departed for Tivoli all at once so that there was no one in the city who could play at the sacrifices. **123** The reverence for God in this affair seized the Senate, and people sent ambassadors to Tivoli to ask that the tibia players return to Rome, as Livy believed; [a3v] when they refused, and the people of Tibur were unable to force them out, the ambassadors invited them on an established day to a banquet to be celebrated with song, and they got them drugged with wine—for which this race is avid. **124** In this condition, they threw them, bound together, into carts and deported them to Rome. **125** They did not become conscious [of this deportation] until the carts had been left on the Forum, [and] daylight oppressed them as they were completely hungover, until they, whose duty it is to decorate [the service] with song, gave in to these persuasions of the populace to stay. **126** Thus were they granted this solemn license to stroll about the city, **127** and the right of eating at public expense was restored to those who sang with the religious worship. **128** Ovid sings about these musicians in his verses of the *Fasti*:

> **129** Add that he ordered that there be only ten artists in an aedile's procession, who should turn out at funerals; they exchange their city for exile and recede to Tibur, and for a certain time their exile was Tibur. **130** The hollow tibia is missed on stage, and it is missed at the altars; **131** no funeral dirge leads final choruses.

132 In like manner, as this art was suited to sacred rites in temples and also to solemnities, so also in all cheerful and more mournful events. **133** As in divine worship the hymns, so also in weddings the wedding-songs and in drinking parties the lyre or kithara were passed around and a convivial type of song was prescribed for those individuals who reclined at table. **134** Among

120 Strab. *Geogr.* 10.3.10; Pl. *Laws* 8.803.
121 Plut. *De mus.* 1136B.
122–127 Livy *Op. om.* 9.30.5.
129–131 Ov. *Fast.* 6.663–667.

the Romans it was the custom at the banquets to recite to the tibiae the noble deeds of distinguished people, expressed in song, so that they gave youth a more eager inclination to imitate them, as the Muse of [Virgil] Maro sang at the banquets of Dido.

> **135** After other leading men, the long-haired Iopetos sounded on his golden kithara what the greatest Atlas teaches. **136** He sings about the labors of the revolving moon and sun.

137 We perceive that this feast has also been celebrated until our times. **138** Likewise, the songs of mourning expressed laments at funerals with tibiae, **138a** as the poem *Fasti* of Ovid relates:

> **139** In the time of our old ancestry there was great use of tibia-players, and this was always in great honor. **140** The tibia sounded in temples; **141** the tibia played at games; **142** the tibia played at sad funerals.

143 Aristoxenus hands down in the first [book] of his *Music* that Olympus first played the aulos at the burial of the Python in the Lydian mode.

144 Song incites rowers to sustain whatever work there be. **145** Song also delights and excites men. **146** It moves the affections, **147** soothes violent passion, **148** tames forest animals, and enables those engaged and tired by laborious work easily to bear it. **149** The melody of the voice consoles the fatigue of individual jobs. **150** It softens the souls of infants, **150a** the young, and the aged. **151** Hence with one "modulation" of the voice (if I may use the word) we seek to influence judges and leaders, with another we ask for mercy. **152** In like manner, as we have already said, music provokes beasts and serpents, birds and dolphins to perceive the pleasant sound of melody.

153 Of how much dignity the discipline of music possesses, the fact that it has the gods themselves for its originators is sufficient evidence. **154** Indeed I speak about the ancient, full-grown, well-constituted, and admirable music, for what was introduced afterward was delicate and petulant, and how much it was to be condemned and valued as worthless [a4r] **154a** Plato teaches in the third [book] of his *Republic*. **155** But our religion does not cease to please the

135–136 Virg. *Aen*. 1.740–742.
137 Ov. *Fast*. 6.651.
138 Ov. *Fast*. 6.652.
138a–142 Ov. *Fast*. 6.652–654.
143 Aristox. *Harm*. 1.
144 Quint. *Ist*. 1.10.16; Virg. *Aen*. 5.177; Ov. *Met*. 3.618 ff; Hor. *Sat*. 1.15.15 ff; Mart. *Spect*. 4.64.21 ff (cf. Wille, *Musica Romana,* p. 123); Cens. *DN* 12.3; Isid. *Etym*. 3.16,2.3; Varro *De mus*. (lost) (cf. Wille, *Musica Romana,* p. 414).
150 Boet. *De inst. mus*. 1.1.186.13.
154a Pl. *Resp*. 3.9–10 condensed.

triune God in sacred buildings, hymns, psalms, antiphons, canticles, and musical instruments, and also in incessant praises through the interceding of other kinds [of instruments], since according to the statutes of the Fathers, [music] provokes the souls to repentance by devoutly disposing the hearts of the hearers and performers. **156** Surely songs and lascivious tunes are prohibited in divine worship, as they incite people to lust and lead the mind away from divine contemplation, especially if the people are guided by pleasure. **157** Therefore, after the temple had been constructed and dedicated, Solomon rightly prayed that whoever entered, crying to the Lord for his sins or whatever troubles, would be heard on those for which he justly entreated.

158 David sang praises to God in the house of the Lord with the kithara and other musical instruments together with singers. **159** Thence it was written and celebrated:

> **160** Praise the Lord with the sound of the trumpet and with psaltery, and kithara, and with string instruments and organ and jubilating cymbals.

161 A sacred law established that the singers properly manage the songs of the sacrifice in front of the altar, [the law] laying divine praises upon all souls.

162 But of those celestial spirits whom Socrates named Sirens in Plato's *Republic,* **162a** the mode is not one and the same that all the blessed themselves and all the heavens in which they reside are said to sing; **163** but in proportion to the diversity [of the spirits] the harmony is as diverse as it is concordant: for the more [the harmonies], the greater and more sweetly they resound. **164** These spirits are seated above according to their orders and their congruence with the spheres, so that the force of every harmony redounds, **164a** as Giorgio Anselmi proposed in the first [book] of his *On Music.* **165** To be sure the Siren is recognized as a singing god, but more properly [the Sirens] are seen as spirits that never cease to sing, as spheres never cease to move.

166 But our theologians better call these spirits "angels," and they distinguish them into nine orders and assign to each one its own function and order. **167** Indeed, those whom they simply call angels are messengers of the divine will on earth; **168** they say that they inhabit the orbit of the elements as a protection to man and that they are witnesses to all things we say and do. **169** They call the second order, presiding over the lunar orbit, Archangels, whose

157 Kings 1. 9.1–3; Chronicles 2. 7.11–12.
158 This statement represents a concatenation of a number of biblical passages from Psalms 27.6; 108.3; 144.9.
160 Psalm 150.3–4.
162 Ans. *De mus.* 1.157; Pl. *Resp.* 10.617b.
162a–164 Ans. *De mus.* 1.156.
165–192, 193–194 Ans. *De mus.* 1.157–168.

function is to be special messengers when the divine will has preconceived what is beyond the [normal] course of nature. **170** The third order is called the Virtues; through them God proclaims miracles and omens to the world, which are signs of coming evil. **171** These Virtues reside in the sphere of Mercury. **172** They call the fourth order the Powers, whose seat in heaven is Venus. **173** These angels control the unclean and wicked spirits, so that they do not have the strength to do harm to the universe and to human nature, which they always watch for as much as they choose. **174** The fifth order is of those who are called Principalities; **175** their seat is the sphere of the sun, **176** but they preside over the multitude of the angels whose function it is to assist the almighty God and the leaders of the earth as much as they have the strength to [a4v] aid the right government of the kingdom. **177** They call the sixth order Dominations; **178** they are the inhabitants of the sphere of Mars. **179** Indeed, they are the leaders of the multitude of armed angels, whom the rank and file obey just as soldiers obey their generals. **180** Moreover, they tame the unjustly fighting people on earth, and they are helpers of righteous men. **181** They call Thrones the seventh order, who inhabit the orbit of Jupiter and announce decrees, divine laws, and judgments. **182** The eighth order is the Cherubim, by which name they comprehend the multiplicity of wisdom. **183** These angels inhabit the sphere of Saturn, and their multitude, in the vicinity of the highest wisdom, confer all comprehensible knowledge to men. **184** They call the ninth order the Seraphim, that is, "burning," and they exceed the other orders of angels in their wisdom, power, blessedness, and joyfulness. **185** The eighth sphere grants a seat to these angels to whom a more intense glow is conceded by means of ampler participation in the divine light of love, joyfulness, and perfection. **186** They say that they veil the divine light, because those spirits and souls placed below them are able only a little to endure the divine appearance, and so they, as it were, keep them from it. **187** Indeed, "angel" is the name of their function, not of their nature. **188** They are quick and altogether indefatigable, running different ways as necessary in an imperceptible amount of time, and for that reason they are furnished with wings through the license of painters.

189 No angel is assigned to inhabit this [ninth] sphere, for it is the place of the eternal God, to whom there is no access by any creature below him. **190** The melody of the other orbits below is contained in this orbit, which completely surpasses every [other] harmony. **191** For this reason, certain of the poets placed Urania there, the one of the nine Muses—whom [the poets] fancied to be the daughters of Jupiter and Memory—who embraces and surpasses all sweetness of speech. **192** Indeed, the theologians assert that the tenth heaven belongs to the highest God, that the ninth seat is inhabited by no one, **192a** as Giorgio Anselmi himself indicated. **193** For below the highest

God, nothing is either worthy or able to come before his divine and ineffable majesty, and nothing at all has the strength to endure such surpassing vigor or conceive it through cognition. **194** Surely all people are to this one as the eyes of a raven blinking at the light.

195 Actually nothing must be valued as either greater or more worthy to the intellect than this celestial, angelic, and almost divine harmony, whose pitches, even if the human ear shuns to hear them. I believe nevertheless that those spirits incessantly jubilate to the highest:

196 Holy, holy, holy, Lord God of Hosts,

196a as Saint Ambrose never ceases to repeat and the most holy Pope Gregory and our holy Mother Church never cease to teach. **197** From there, our mind is able to wish for nothing sweeter, nothing more pleasant than a return to heaven, to be dissolved from these bonds of the flesh and [a5r] to be associated with these most consecrated spirits.

198 This divine and celestial science, which alone delights the highest creator and the celestial spirits, comes forth to all inferior nature, so that if we take our argument from human nature, then music is of use and profit in the several periods of life, and Aristotle teaches **198a** that it brings natural delight to these and is a friend of and in agreement with the individual moral characters. **199** Indeed, we read that Chrysippus assigned a certain proper song to nurses who were employed in nursing infants to suppress their crying.

200 Socrates and Plato, and the Pythagoreans, too, attributing moral preparation to music, ordained that by common law adults and youths and young men and women be educated in musical science, not for the sake of inciting to lasciviousness, by which this science becomes cheapened, but for moderating the motions of the soul under the rule and reason. **201** As not every voice but only that which agrees well with the melody of the sound has value, so not all motions but only those which agree with reason belong to the right harmony of life. **202** God granted us voice and hearing for the sake of this, as Plato seems to assert in the *Timaeus,* for speech pertains and conduces very much to these things. **203** Truly, every use of the musical voice is bestowed by the grace of harmony, **204** and harmony, which possesses motions congruent and inborn to the course of our soul, is useful to a man who

196 Isaiah 6.3.
198a Arist. *Pol.* 1339.17–25.
199 Quint. *Ist.* 1.10.32 (as cited in: Günther Wille, "The Use of Rhythm and Music in the Pedagogical Therapy of Classical Antiquity," in *Orff-Yearbook* 1962, p. 50; also Wille, *Musica Romana,* p. 148).
200 Pl. *Resp.* 399, 410–411.
202–206 Pl. *Ti.* 47 c-e (trans. Fic.) 533 F–G.

prudently uses the arts not for pleasure without reason, as it now seems. **205** Rather harmony was given by the Muses that by it we might calm every dissonant motion of the soul and lead it back to harmony in agreement with itself. **206** Rhythm, too, seems to have been bestowed for this purpose, so that we may very suitably temper the immoderate and graceless condition within us. **207** Indeed, since the nature of boys is restless and in everything desirous of delights and cannot sustain any severe discipline, Plato himself justly orders that boys be educated in the medium of respectable music which prefers to pleasure the pathways of this most common virtue. **208** True, there sometimes are older people among whom certain decorous delights should be employed as solace for a laborious life. **209** So in youth comforts must be provided which have the power to delight an honest old age. **210** Therefore Plato restrained the trifold urge—that is, for food, drinking, and sex—by a triple bridle, in order that they not perish in violence; **211** this bridle consists of fear, law, and true reasoning. **211a** Thus by moderate use of honorable games from childhood on through honorable comforts they would be recalled from disgraceful things and gradually incited to serious studies. **212** Therefore he wished them [to spend their leisure] in exercise of the body and to employ gymnastic games such as dancing and wrestling.

213 Some dancing imitates the words of the Muse and observes her magnificence and liberty; **214** other dancing, by its grace, aptly flexes and stretches single elements of good carriage, lightness, and form of the body itself, and of its parts and members, [a5r] just as harmonious motion is sufficiently diffused as it follows the complete order of the dance. **215** Thus Dionysius of Halicarnassus writes in the second book about the Salinian priests: **216** by the music of the aulos they are moved to arms in enoplian motions, as appropriate, sometimes together, other times in turn, **216a** singing certain ancestral songs, dancing the three-step war dance and solemn leaps, together with choral dances. **217** If one may conjecture from the reports of the men of former time, the Curetes were among the first to establish the dance, the drill with arms, and the clatter made by the swords on shields. **218** In truth, these things are employed in proper wrestling from the neck to the hands, by the motion of the sides with eagerness for victory and an honorable attitude; **219** and they lead men to health and strength, since he judged them useful in all things and

207–211 Pl. *Resp.* 439 a–e (trans. Fic.).
211a–214 Pl. *Leg.* 673 (trans. Fic.).
215–216 Dion. Hal. *Ant. Rom.* 2.70.
216a Livy *Op. om.* 1.20.4 (as cited in Wille, *Musica Romana,* p. 43); Dion. Hal. *Ant. Rom.* 2.70.
217–219 Dion. Hal. *Ant. Rom.* 2.70.

not at all to be neglected. **220** This is what Marcus Tullius [Cicero], too, reports as the opinion of Plato himself in the book on the *Laws* with these words. **221** As there already are public games, divided among the theater and the circus, the athletic contests—running, boxing, wrestling, and horse races—are carried out to a sure victory in the circus; for the song of the voice, of stringed instruments, and of the aulos the place is the theater, provided that these entertainments are moderate, as is prescribed in the law.

222 I agree with Plato that nothing flows into tender and soft souls so easily as the various sounds of singing. **223** It can hardly be stated how great its power in every respect is, **224** for singing incites the feeble and makes the lively faint, and at one time it relaxes, at another time tenses souls. **225** To preserve the ancient manner of singing was of importance in Greece to many cities, whose customs, having slipped into weakness, were altered along with their songs, either perverted by this pleasantness and by corruption, as certain people believe, or because their severity of custom had vanished on account of other vices. **226** Thereupon, in the transformed ears and souls there also was a place for this change, on account of which the wisest and by far most learned man of Greece, Plato, strongly feared a decline. **227** In fact, he denies that musical principles can be altered without a change in the common laws. **228** Truly, among the Athenians the use of the aulos was in style and was beneficial, as Horace mentions in this passage from his *Art of Poetry*:

> **229** The aulos was not, as now, joined together with bronze alloy and a rival to the trumpet, but was thin, and simple with a small bore, [and] was useful to accompany and lead the chorus, etc.

230 But when Alcibiades as a boy was sent to the aulete Antiginides to be taught, he, ashamed of deforming his mouth, threw away and broke the auloi he had brought to his mouth and blown into. **231** When this fact had spread abroad, at that time by the consensus of all Athenians the discipline of singing to the auloi was stopped. **232** I believe, though, that the discord of sounds in song is more disgraceful than the deformation of the mouth.

233 It is said that many cities were founded in Greece, and laws and decrees instituted by musical reason were recited to the lyre or auloi. **234** But in this

220–221 Cic. *Leg*. 2.38.
222 See the annotation to Kreyszig, "Franchino Gaffurio's *Theorica muusice*," vol. 3 (Edition) (*Theorica* 1.1.222).
222–224 Cic. *Leg*. 2.38; Pl. *Resp*. 401 d–e.
225–227 Cic. *Leg*. 2.38–39; Pl. *Resp*. 424 c.
228–229 Hor. *Ars P.* 202–204.
230–231 Plut. *De vi Alc.*
232 Cap. *De nup*. 9.492.8–10.
234 Cap. *De nup*. 9.492.8–10; Varro *De mus*. (lost) (cf. Wille, *Musica Romana,* p. 415).

our time the leaders and the Senate of the state are accustomed to announce by tuba, horns, and so forth laws and decrees which must be observed. **235** Therefore, among the ancients, this music [a6r] was a discipline for use either in divine cult or in the teaching of youth or in the maintaining of the state and community, and without any apparatus of the theater, which did not yet exist.

236 Music was used in churches, where people were eager both to venerate the divine power and to praise good men. **237** Hence Augustine, in the second [book] of the *City of God,* reports that Scipio said that from stringed instruments and auloi and in song itself and voices, a certain harmony must be achieved from the separate sounds. **238** The trained ear cannot tolerate this harmony to be changed or to be made disharmonious granted that this harmony is nevertheless made concordant and congruent by the moderation of highly dissimilar voices. **239** Thus from the highest, middle, and lowest orders intermingled, as with sounds, the state is moderated by reason and harmonizes by the consensus of very dissimilar [persons], and that which is called harmony by musicians in song is what is concord in a city, the tightest and best bond of safety of all in a state, and it can by no means exist without justice.

240 The wisest poets have acknowledged that music is a friend of liberal disciplines and necessarily agrees with them. **241** Archytas and Aristoxenus stated that grammar is subject to music. **242** Eupolis taught [both] music and literature, **243** and Maricas, who is Hyperbolus, acknowledged that he learned letters from musicians. **244** Aristophanes pointed out in more than one book that in former times boys were taught in this manner. **245** Moreover, the logicians adored music, **245a** thereby claiming voice and sound as their subject. **246** Marcus Tullius [Cicero] did not neglect music; **247** certainly nothing contributes more to orators than the care of the voice, so pertinent to music. **248** When Gaius Gracchus gave a speech to the Roman people in his tribuneship, he kept behind him a slave, experienced in the musical art, who secretly shaped the modes of the orator's delivery with an ivory pipe, exciting them when they were too languid or restraining them when they were violently moved beyond what was just, because the heat and impetus of his delivery did not permit him to be an attentive estimator of his own tempera-

237–239 August. *De civ. D.* 2.21.

244 Ar. *Nub.* 2.647ff., 961ff.; Ar. *Ran.* 2.1500ff.

246 Cic. *De Or.* 3.174ff.

248 Val. Max. 8.10.1; Cic. *De Or.* 3.60.225; Quint. *Ist.* 1.10.27–28; Gell. *NA* 1.11.10–13, 15 (cf. Wille, *Musica Romana,* pp. 453–454). This account is also reported in an undated anonymous letter to Giovanni del Lago (c. 1490–after 1543), preserved in Manuscript Paris, Bibliothèque Nationale, ms. it. 1110, folios 5r–6v (cf. Bonnie J. Blackburn, Edward E. Lowinsky, and Clement A. Miller, eds., *A Correspondence of Renaissance Musicians* (Oxford: Clarendon Press, 1991), pp. 921, 925).

ment. **249** The principal of a gymnasium reproved Carneades of Cyrene, who was reading in a public game with a louder voice than was proper, as he had not imposed any moderation of speech on his voice.

250 The arithmeticians Theophrastus of Crete **250a** and Nicomachus systematized music. **251** The geometricians Theodorus, the Cyrenaic, and Archytas made music their profession. **252** The astronomers Pythagoras and Ptolemy disclosed musical consonances by sonorous numbers. **253** The theologian Plato made the very pleasant harmony of the universe depend on the ear and on reasoning. **254** Finally, all people in Greece learned to sing, and one who did not know how was believed to be lacking in erudition. **255** Indeed, they say that Themistocles was of uncultured character, for when a lyre was offered to him at a banquet, he indicated himself unskilled. **256** That no scientific discipline can be understood from proposed imitations and examples without music, and that nothing can be made or exist without music itself, Isidore has laid down in his *Etymologies*. Moreover, Hrabanus [Maurus] in a treatise he devotes to this discipline asserts this same thing [a7r]. **257** It is pleasure to celebrate succinctly here the memory of those most sagacious authors who left musical monuments to posterity. **258** Indeed, Pythagoras of Samos, who first claimed the title of philosopher for himself, is said to have devoted himself within the mathematical sciences foremost to music and geometry and to have discovered the rule which is derived from one string. **259** Socrates, already advanced in age, learned to play on string instruments, believing that one must by no means feel ashamed to learn what one does not know. **260** For that reason he handed down musical precepts for the instruction of youths. **261** Bion, as Eratosthenes mentioned, was the first of all who, adorning philosophy by diverse embellishment of speech, had music and geometry for diversion. **262** Theodorus the Cyrenaic, under whom Plato

249 Diog. Laert. *Vitae* 4–63 (kindly identified by Dr. Holford-Strevens).
250 Theophr.'s *De mus.*, a work no longer extant, is quoted by Porph. in his *Harm.*; see Lukas Richter, "Porphyry," *New Grove* 15 (1980): 123. Excerpts from Theophr.'s writings are also included in Pliny *HN* 2.95.209; Varro *Rust.* 3.17.4; and Cap. *De nup.* 9.493.3,20 (cf. Wille, *Musica Romana,* p. 416). The remark of the latter author is cited in *Theorica* 1.1.66–67. Theophr.'s comment on the healing power of music is quoted in Gell. *NA* 4.13.2 as well as in *Theorica* 1.1.92. Theophr.'s medical practices are also mentioned in Mar. Vict. *Ars gram.* (cf. Wille, *Musica Romana,* p. 602). On Nicom., see below 1.1.277.
250a Nicom. *Man., Intro. to Arith.*
251 Theodorus's musical connection cannot be documented. None of his writings are extant. Arch.'s *De math.* is quoted by Porph. in his *Harm.*; see Lukas Richter, "Porphyry": 123.
252 On Ptol., see below 1.1.276.
255 Quint. *Ist.* 1.10.19 (cf. Wille, *Musica Romana,* p. 452).
256 On a more specific identification of Isid. *Etym.*, see below 1.1.282; further on Hrab.'s writings, see 1.1.282.
259–260 Quint. *Ist.* 1.10.3–14.
262 On Theodorous the Cyrenaic, see above 1.1.251.

studied, compiled a small book on geometry and music. **263** Theodorus, another musician, wrote a small book on the training of the voice. **264** The Athenian musician Xanthus wrote down rules for harmony. **265** The aulete Dionysodorus handed down rules for playing the auloi. **266** The Pythagorean Philolaus, having very accurately investigated the musical elements—that is, the tone and the semitones—was of the opinion that all things were necessarily produced from harmony. **267** Because Democritus of Milet wrote on almost all matters and disciplines, **267a** he also composed many volumes on music. **268** The Pythagorean Archytas of Tarent was the first to measure musical motion by a geometrical description. **269** The Athenian Simon wrote a copious dialogue on music. **270** Moreover, with [Diogenes] Laertius as a witness, Simmias of Thebes left to posterity an elegantly written musical dialogue. **271** Heraclides of Ponticus wrote two books on music. **272** In the *Timaeus* and in the *Laws,* the divine Plato treated whatever comes together by harmonic persuasion in human natures, in celestial bodies and also in the elements. **273** Theophrastus of Crete wrote three books on music and also one on musicians. **274** The Platonic Xenocrates of Calchos investigated that the

263 Diog. Laert. *Vitae* 2.103 (kindly identified by Dr. Holford-Strevens).

264 According to Ath.'s *Deip.* 12.513 A.6, Xanthus is said to have written an *Oresteia* (cf. Michaelides, *The Music of Ancient Greece,* p. 346).

266 Phil., fragm. *De nat.* contains an analysis of the Pythagorean doctrines on music (cf. Michaelides, *The Music of Ancient Greece,* p. 248). An examination of Phil.'s views on music including the division of tone and semitone, is included in Nicom. *Man.* (cf. Michaelides, *The Music of Ancient Greece,* p. 248).

267 For Democr.'s writings on ethics, physics, mathematics, philology, and agriculture, see Hermann Diels, *Die Fragmente der Vorsokratiker,* 3 vols., 6th ed. (Greek and German), ed. Walther Kranz (Berlin-Grunewald: Weidmannsche Verlagsbuchhandlung, 1951–1952, 2:130–149. Further on Democr.'s discussion of physics, see Cyril Bailey, *The Greek Atomists and Epicurus* (New York: Russell & Russell, 1964), p. 114.

267a Democr. *On Rhyth.* (cf. Diels, *Die Fragmente der Vorsokratiker,* 2; 145). Democr.'s contributions are mentioned in Gell. *NA* 5.13.3–8. Gell.'s remarks are cited in *Theorica* 2.4.24.

268 On Arch., see above 1.1.251.

269 Simon's dialogue, no longer extant, is mentioned in Diog. Laert. *Vitae* 2.14.

270 Simmias's dialogue on music is mentioned in Diog. Laert. *Vitae* 1.2.

271 Heraclid.'s collection is quoted in Plut. *De mus.* 1131F. Furthermore, Porph. cites Heraclid. twice, that is, in connection with a remark about Pythagoras and Xenocrates from his "peri touton en te Mousike eisagoge" (Ingemar Düring, *Ptolelaios und Porphyrius über die Musik,* p. 30, lines 1–8), the other a note (unidentified by Porph.) saying "ex Heraclide allatum esse crediderim" (Düring, p. 32, line 23 and p. 33, line 4). I thank Professor Palisca for locating these references in Porph. Heraclid.'s *On Music* is cited in Ath. *Deip.* 14.624 C.19 (cf. Michaelides, *The Music of Ancient Greece,* p. 135). Gaf. (in *Theorica* 1.1.3, 3b) cites a passage from Heraclid.'s *Coll.,* in Plut.'s *De mus.*

272 Pl. *Ti.* 47c.1.7–e.1.2; 88c.2.4–5, 80b.2.4–8; Pl. *Leg.* 802c.1.6–d.1.6.

273 On Theophr. see above 1.1.250.

274 Reference to Xenocrates of Calchos is provided in Macrob. *Somn. Scip.* 1.14, in Val.

mathematical sciences and especially music confer not a little to philosophy. **275** Archelaus, the teacher of Socrates, and the Athenian Xenophon, whom they called the Attican Muse because of the sweetness of his eloquence and his incredible facility, and also Demosthenes, Anaxagoras, Damon, and Porphyry dealt with the excellence of music in their books. **276** Ptolemy wrote a copious book on musical art. **277** But Aristoxenus and Nicomachus, the Athenian Aristo, and Albinus transmitted many volumes on consonances. **278** Aristotle wrote one book on music **278a** and another on Pythian music, **278b** and he also included problems and musical elements in many books. **279**

Max. 9.10 (cf. *Lemprière,* p. 721) as well as in Cap. *De nup.* 9.493.5. The remark of the latter author is cited in *Theorica* 1.1.85.

275 Archelaus's contribution is mentioned in Cic. *Tusc.* 5 and in August. *De civ. D.* 8 (cf. *Lemprière,* p. 74). On Xenophon of Athens, see Quint. *Ist.* 10.2 (cf. *Lemprière,* p. 725). Quint. *Ist.* 12.2.22. On Dem.'s speaking with pebbles in his mouth, see Quint. *Ist.* 11.3.54 (cf. Wille, *Musica Romana,* p. 485). Gaf. includes an implicit reference to this account in *Theorica* 2.6.23. On Anaxagoras, see: *Lemprière,* p. 50. Dam., fragments of his *Areo.* on rhythm and ethical value of music. Damon's view on music are communicated by several authors, among them, Pl. (*Resp.* 4.424 C); Ath. (*Deip.* 14.624 C.25); and Aristid. Quint. (*De mus.* 2.14) (cf. Michaelides, *The Music of Ancient Greece,* pp. 71–72). Gaf. (*Theorica* 1.1.82) mentions Dam. in a passage from Cap. *De nup.* 9.492.16. Porph. *Harm.* Gaf. includes a further reference to Porph.'s treatise in *Theorica* 1.2.36b.

276 Ptol. *Harm.*

277 Aristox. *Harm.*; Aristox., fragm. on *Harm.* 4.667 (cf. R. P. Winnington-Ingram, "Aristoxenos," *New Grove* 1:591). Nicom. *Man.*; Nicom., lost *opus maior* (mentioned in *Man.* 242.11; cf. Wille, *Musica Romana,* p. 662). The only known reference to Aristo of Athens is given in Pl. *Leg.* 670 b.1 (cf. Wille, *Musica Romana,* p. 462). Albinus, lost treatise on music (cf. Carl von Jan, "Der Musikschriftsteller Albinus," *Philologus* 56 [1897]:163–166). Albinus's treatise is cited in Boet. *De inst. mus.* 1.12.199.14 and 1.26.218.22. Albinus is further mentioned in Cassiod. *De mus.* 10.1212BC (cf. Wille, *Musica Romana,* p. 707). Gaf. includes references to Albinus in *Theorica* 2.2.116a; 5.1.105.

278 Arist. lost treatise on music (cf. Warren Anderson, "Aristotle," *New Grove* 1:589). This treatise is mentioned in the catalogue of Arist.'s writings, but there is no mention of this work or any other record in his extant writings (cf. Ulrich Fleischer, "Aristoteles," *MGG* 1 (1949–1951), col. 632). Arist. further elaborated on music and on elements relevant to music in the following treatises: *De an., Metaph.; Eth. Nic., Ph., Cael., Pol.,* and *Top.* For specific references concerning the discussion of music in these treatises, see: Warren Anderson, "Aristotle," *New Grove* 1:587–591, also note to 1.1.278b.

278a There exists no record of a work of Arist. on Pythian music (cf. Fleischer, "Aristoteles," cols. 632–633; Anderson, "Aristotle": 587–591).

278b For example, in *De an.* 408a.2.5–10; *Metaph.* 958b.1.32–986a.1.2; 1090a.2.20–23; 1093a.1.28–1093b.1.4;1018b.1.29; *Eth. Nic.* 1103a.1.14–1106a.1.24; *Ph.* 1.5; *Cael.* 290b.2.21–23; *Pol.* 1254a.2.32–33; 1290a.2.19–22; 1340a.1.15; 1170a.2.6–11; 1339a.1.11–1342b.1.34; 1338a.2.22–23; *Top.* 139b.2.37–38; *Poet.* 1448b.2.20–21; *Pr.* (original lost, preserved copy compiled from the writings of Theophr.; cf. Fleischer, "Aristoteles," col. 633).

279 Epic. *Spic.* (cf. Hermann Usener, ed., *Epicurea* [Leipzig, 1887]). Epic.'s contributions are mentioned in Lucr. *Op. omn.* 3–6 (cf. Wille, *Musica Romana,* p. 420 ff.) and in Gell. *NA* 5.15.3–8. The remark of the latter author is cited in 2.4.24.

Likewise, Epicurus treated music at length. **280** Authors who followed were Bryennius, Aristides Quintilianus, Bacchius Senior, and Plutarch, the more recent among the Greek authors, who wrote copious volumes on this topic for posterity.

281 Finally, [Latin] antiquity honored those authors whom musical reasoning did not desert. **282** Chalcidius and Macrobius, Censorinus and Cassiodorus, Isidore and Hrabanus, Fabius Quintilian and also the architect Vitruvius—very eminent men—are known to have venerated the musical discipline with the greatest assiduity [a7r]. **283** Severinus Boethius, whose genius, we read, no one after him approached, compiled most accurately what he thought necessary to musical speculation in a five-part volume in a Latin interpretation. **284** His subsequent interpreters, Ugolino of Orvieto, Giorgio Anselmi of Parma, and others, explicate him with many famous and almost infinite works. **285** But Guido of Arezzo revealed whatever was regarded as obscure and unknown in a simpler form and instituted a common method of making melody, which, as Guido himself attested in a letter to Michael of Pomposa, Pope John [XIX] approved by apostolic authority. **286** And at that time, they say, Abbot Odo published his musical dialogue. **287**

280 Bryen. *Harm.*; Aristid. Quint. *De mus.*; Bac. *Intro.* Gaf. had commissioned a translation of this last work from Bur. The date of completion of this translation is not known, for no copy of it has survived (cf. Palisca, *Humanism,* p. 192). Plut. *De mus.*; Plut. *De pro. in Ti.* (cf. Michaelides, *The Music of Ancient Greece,* p. 261).

282 Chalcid. *in Tim.*; Macrob. *Somn. Scip.* 2.1–4. Gaf. includes reference to this treatise in 1.8.18; Cens. *DN* 10, 12, 13; see also the anonymous *Fragm. Cens. de mus.* (cf. Warren Anderson, "Censorinus," *New Grove* 4: 56; Wille, *Musica Romana,* p. 598). Cassiod. *De inst. div.* 2.5 ("De musica"); *Expo. in psalt.*; *Var.* 2.40 (edited in 537 A.D.). Gaf. quotes extensively from Cassiod.'s letter to Boet. below 5.8.46–52. Isid. *Etym.* 3.15–23; *De eccl.* 1.3–7, 9, 13–15; 2.12. Hrab. *De cler.* 3.24; 1.32–33; 2.1–9; 2.47–51; *De univ.* 18.4 (cf. Helmut Giesel, *Studien zur Symbolik der Musikinstrumente im Schrifttum der alten und mittelalterlichen Kirche (von den Anfängen bis zum 13. Jahrhundert)* vol. 94 of *Kölner Beiträge zur Musikforschung* ed. by Heinrich Hüschen (Regensburg: Gustav Bosse, 1978), p. 87 ff.); *De sac.* 18–19; Quint. *Ist.* 1.10; Vitr. *De arch* 5.

283 Boet. *De inst. mus.*

284 Ugol. *Decl.*; Ans. *De mus.* Gaf. quotes extensively from this work in 1.1.162–194.

285 Gaf. here refers to an antiphoner, now lost, probably using staff lines for F and C. Guido explained the notation in a letter, entitled *Epist.*, to his friend Brother Michael of Pomposa. In *Theorica* 5.6, Gaf. quotes from this letter as well as from the prologue of the lost antiphoner; the same chapter of the *Theorica* also includes quotations from two other treatises of Guido, namely, the *Reg.*, the poetic form of the Prologue to the Antiphoner, and the *Microl.* (cf. Claude V. Palisca, "Guido of Arezzo," *New Grove* 7:804–807).

286 An anonymous musical dialogue wrongly attributed to Abbot Odo of Arezzo. Gaf. reports this wrong attribution, as circulated in numerous sources of the dialogue (cf. Michel Huglo, "L'auteur du 'Dialogue sur la musique' attribué à Odon," *Revue de Musicologie* 55 (1969): 119–171).

287 Johannes Carthusensis, Johannes Mantuanus, Johannes Gallicus. Gall. wrote three musical treatises, entitled *Rit. can.*, *Praef.*, and *Tacita.* (cf. Cecil Adkins, "Legrense, Johannes,"

Having imitated Odo's and Guido's tradition, Johannes Cartusinus wrote a very useful small book on the art of music for clergymen. **288** Augustine, on the other hand, produced six books on music, **289** while Saint Ambrose, our protector [that is, of the Milanese] and also Gelasius and [Pope] Gregory instituted that graduals, tracts, and alleluias be sung during mass. **290** And most of the other very eminent men are known to have added certain other [things]. **291** Moreover, Ambrose imitated in many [aspects] the rite of the Greeks, which Basil and the other most holy men had instituted. **292** Indeed, a new order of singing hymns was brought to a flowering at first by Bishop Hilary of Poitiers. **293** And after him Ambrose is celebrated, and from that time these hymns are called Ambrosian hymns because they began to be sung frequently at his time in the churches of Milan. **294** Indeed, whatever songs after these were composed in the praise of God in the western churches, they are called hymns. **295** Finally, Ambrose and Gregory instituted that in the divine offices almost all songs are to be sung by musical voices, but not in composite and florid harmony, nor dissonant as is the vogue among our Ambrosians in the burials and vigils for the martyrs, but of a more simple and plain harmony. **296** The priests who succeeded them introduced this very sweet and consonant harmony in the divine services, imitating David, who, when he wished to glorify the cult of God with more solemnity, we read, introduced singers before the Ark of the Covenant of the Lord and, therefore, singing with musical instruments and melodious voices. **297** These musicians softened by harmonious sweetness the souls of those who stood with them and soon made them more attentive to divine praises. **298** Now the brevity of time urges me to omit the most eminent musicians and singers of our time, whom our generation in its praises celebrates as by far more skilled than their predecessors. **299** But we believe that the only part of this very delightful science that pertains to us is that which establishes character; **300** indeed, this makes men glorious and eternal, and hands down immortals to posterity.

New Grove 10:615). Though none of the three treatises are specifically written for clergymen, they all include sections in dialogue form on the musical interpretation of plainchant, illustrated using specific examples from the liturgy, as well as interpretations of the modal system, the psalm tones, and the solmisation—topics of vital interest to clergymen.

288 August. *De mus*. Additional references to August.'s treatise are included in 1.4.16–21, 23–24; 1.5.7a–11; 1.5.12–18; 1.6.14, 14b, 14c, 15–16a, 18–19, 2.2.47–72, 73–107; 2.8.1–6, 10a, 13–35; 3.3.47–54. On the identification of 2.3.35 with August. *De mus.*, see annotation to the translation.

289 Concerning the introduction of singing these portions of the mass, see James W. McKinnon, "Ambrose," *New Grove* 1:314.

292 Hil. *Tract*. Cf. Heinrich Hüschen, "Hilarius von Poitiers," *MGG* 6 (1957), col. 379).

296 1 Chronicles 15.27–28.

Chapter **Two**
Musica mundana

1 Several philosophers have written that music is produced [in response to] the proper arrangement of the rapid whirling motion of heavenly bodies, **1a** the seasons, and the elements. **2** The Pythagoreans explained that this [music arose] from the sounds and consonances of the heavenly orbits, produced by rapid whirling motion or collision. **3** Indeed, these celestial bodies and the corresponding elements, mixed together simultaneously by a certain combination of consonance, completed this whole universe. **4** This universe is called "mundus" by the philosophers, because it is always in motion. **5** They do not believe that he—the highest and best creator and builder of all things—perfected such a great and so very significant machine to be motionless in its useless and inapt taciturnity, that anything was missing to complete this most ornate of all works, so that the work would resound harmoniously and bring forth many agreements of sounds. **6** Accordingly, [the philosophers] are of the opinion that thus heaven is put in motion, and those happy troops of heavenly spirits, and our souls, when they retire to that place, sing wonderful or harmonious songs. **7** Therefore it cannot be the case **7a** that such a rapid machine of the heavens should move in a silent and quiet course, **7b** though its sound does not reach all the way to our ears. **8** This is the opinion of the Pythagoreans. **8a** For sound is inborn from the beginning of our nativity, as it appears in hammerers who scarcely perceive their own noise, **8b** and also for many other reasons it is necessary that this happen—[an opinion] **8c** which Aristotle seems to reject in [his] second [book] on *The Heavens and the Universe*.

9 Yet it cannot be that such very rapid motions of so great bodies excite no

1a For example, Boet. *De inst. mus.* 1.2.187.25–26; see also Gaf. *De harm.* 4.19.

4 Cf. Isid. *Etym.* 3.29;13.1 (cf. James Haar, "Musica Mundana: Variations on a Pythagorean Theme," Ph.D. diss., Harvard University, 1960; p. 365).

5–6 Ans. *De mus.* 1.148.

7–7a Boet. *De inst. mus.* 1.2.187.26–28.

7–7b Cf. Arist. *Cael.* 2.9; Arist. Quint. *De mus.* 3.20; Nicom. *Man.* 3; Boet. *De inst. mus.* 1.3; Macrob. *Somn. Scip.* 2.4.14. This information was kindly communicated to the translator by Professor Mathiesen on 4 February 1992.

8b Boet. *De inst. mus.* 1.2.187.28–29.

8c Arist. *Cael.* 2.9.290B–291A, as cited in Ans. *Dia.* fol. 5 verso. Haar (cf. "Musica mundana", p. 366) notices a certain affinity to a statement in Ans., *Dia.* (cf. Giuseppe Massera, "Un musico parmense: Anselmi Giorgio 'Senior,'" estratto da *Aurea Parma* 39, fasc. IV [1955], 27n).

9–10 Boet. *De inst. mus.* 1.2.187.29–30; 188.1–6; see also August. *De civ. D.* 22 (*PL* 41, col. 791). Concerning Boet.'s familiarity with August.'s works, see Boet. *De uni. Trin.*, Prooemium (cf. *PL* 64, col. 1249). This close affinity between the respective passages in Boet. *De inst. mus.* and August. *De civ. D.* is discussed in Haar, "Musica mundana," p. 169.

sounds at all, especially as the courses of the stars are bound together by such accurate joining-together that nothing known is equally joined together, nothing so well mixed. **10** For some orbits are said to be higher, others lower, and in this manner all are moved by an equal violent motion, so that through the different inequalities a rational order of the courses is introduced. **11** Whence the soul itself, regulating the grand forces of heaven, now through intensity, now through subsidence, confines those divine spirits to a concordant sound, so that the forces accompany in turn all the diversities of the harmonious voices. **12** But the soul, not content with a single tone, changes at one time to the diatessaron, at another time to the diapente, at yet another time to the diapason, and to other and always different consonances. **13** Nor does a single sphere bring forth merely a single *harmonia* continuously but varied pitches, so that these very happy spirits seem sometimes with the sound of their own sphere, sometimes with those that are neighboring—at one time surpassing in the song, at another time following later, and yet at another time following instantly, at another time coming together with them—so that they are believed always to contend very sweetly in a wonderful play of the agreement of sounds [*harmonia*].

14 Therefore it cannot be that the deliberate order of making melody deviates from this heavenly revolution. **15** [Marcus] Tullius [Cicero], indeed, very intelligibly said with the profundity of Plato's own words that "making of melody" delights the Platonic ears. **16** "What is this," I say "and what is this sound, so loud and so agreeable, that fills my ears?" [a8r]. **17** "That is produced," he replied, "by the onward rush and motion of the spheres themselves; the intervals between them, though unequal, being exactly arranged in a fixed proportion."

18 And yet, the soul by which the heavens are revolved and [by which] all living things are animated is itself controlled by certain laws. **19** The numbers by which the bond is recognized among these living beings attest to this—if indeed these numbers are those in which they say that our sensible harmony is sung. **20** To be sure, since the Almighty God deliberately determined to construct this universe in this way, it was necessary that he add the soul as a guide and mover, for bodies are not moved by themselves unless animated. **21** Yet if the soul were incorporeal, it would be able to furnish the world with a body and what is contained in it.

22 The soul of the universe is joined to its mundane body through harmo-

11–13 Ans. *De mus.* 1.149–150. Ans. derives the terminology (*accentus* translated as "intensity"; *succentus* translated as "subsidence") from Chalcid. *in Tim.* 16 (cf. Haar, "Musica mundana," p. 360).
16–17 Cic. *Rep.* 6.18.18–19.
18–21 Ans. *De mus.* 1.134–135.

nious proportion, namely, the duple, triple, quadruple, sesquialter, sesquitertial, and sesquioctaval proportions, something Giorgio Anselmi has examined. **23** Therefore, no doubt, the solid and sounding body of the heavenly machine, woven out of these numbers and proportions, through which it is evident all harmony established, gives forth sounds of amicable sweetness when moved steadily and most swiftly, especially because the soul, which is totally filled with intelligence, moving not accidentally nor by chance but in the right order, moves and governs the universe, the body of the world itself. **24** I noticed that Saint Ambrose mentions this opinion in the second day of his *Hexameron*. **25** But unless there were some harmony in the diversities and contrary qualities of the four elements joining them, as Boethius asserts, **25a** they would hardly agree [enough] to form one body and one mechanism. **26** In fact, all this diversity produces the variety of the seasons and the fruits so that only one integral year emerges. **27** Whence, if you should separate any of the bodies, which provide such a great variety to things, from mind and thought, it is necessary that everything would perish, nor would they preserve anything consonant. **28** Just as in the low steps there emerges this type of pitch, so that the lowness does not descend all the way to silence, **29** and also that type of height is maintained in the high steps, so that the strings which are stretched are not broken by the tenseness of the pitch, but everything is in itself agreeing and everything convenes, **30** so also, even in the music of the world, we discern that nothing is so in excess that it dissolves another [thing] by its own excess.

31 Truly, whatever exists either produces its own fruit or assists others so that they bear fruit, for what the winter restrains, spring lets go, summer dries up, and fall ripens. **32** So the seasons in turn produce their fruit by themselves or bring help to the others so that they produce. **33** Therefore the universe is composed by means of a certain agreement of sounds [*harmonia*], **33a** and heaven itself is turned in the harmonious making of melody, **33b** but the distances of heavenly bodies are estimated solely by ratio, not by sense.

23 Ans. *De mus.* 1.137.
24 Ambr. *Hexam.* 2.5.21; also 2.2.7.
25 Boet. *De inst. mus.* 1.2.188.7–9. Further on the elements, see Pl. *Symp.* 188A; Pl. *Ti.* 32C; Macrob. *Somn. Scip.* 1.5.25.
25a Boet. *De inst. mus.* 1.2.188.9–10.
26 Boet. *De inst. mus.* 1.2.188.10–12. Further on the seasons, see Pl. *Symp.* 188A.
27 Boet. *De inst. mus.* 1.2.188.12–15; see also Ptol. *Harm.* 3.4. Cf. Düring, *Die Harmonielehre des Klaudios Ptolemaios,* p. 117.
28 Cic. *Rep.* 6.18.18–19; Boet. *De inst. mus.* 1.2.188.15–17.
31–32 Boet. *De inst. mus.* 1.2.188.17–25.
33, 33b Macrob. *Somn. Scip.* 2.3.13.
33a Isid. *Etym.* 3.17.

34 Archimedes, as testified by Macrobius, believed that he discovered the number of stades by which the moon is distant from the surface of the earth, Mercury from the moon, Venus from Mercury, the sun from Venus, Mars from the sun, Jupiter from Mars, and Saturn from Jupiter; **35** he also thought that the entire space from the orbit of Saturn all the way to the star-bearing heaven is measured by its ratio. **36** The Platonists, however, have repudiated his measurements because they do not observe the duple and triple intervals, **36a** but [the Platonists] have determined that another distance must be stretched out for it, **36b** which Porphyry introduced in his books. **37** Pythagoras, however, indicated how many stades there are between the earth and the individual stars by a musical analogy. **38** Eratosthenes concluded by geometric reasoning that the maximum circumference of the earth [a8v] is 250,000 stades. **39** Now they understood in this measurement of the universe that the stade was 625 feet; **40** there are other stades differing in length, as for example, the Olympic, which is 600 feet, and similarly the Pythian is 1000 feet. **41** Consequently, Pythagoras believed that from the earth to the moon there are about 126,000 stades and that this is the interval of the tone. **42** The distance from the moon to the planet Mercury, which the Pythagoreans called Stilbon ["twinkler"] corresponds to the larger semitone—**43** from this place to the morning star Venus is almost as far, that is, the smaller semitone. **44** From that place onward to the sun, the distance is little more than merely three times as much, that is, a tone and a smaller semitone. **45** Therefore the sun is distant from the earth three tones and a smaller semitone, which is called the diapente; **46** however, the sun from the moon two tones and a semitone, which is the diatessaron. **47** From the sun to the planet Mars, whose name is Pyrios, there is just as large an interval as from the earth to the moon, and this makes the tone. **48** From that place to the planet Jupiter, which is called Phaetheon, that is, the half [*dimidium*] of this, makes the smaller semitone. **49** From Jupiter all the way to the planet Saturn, which is called Phaenon, the distance is a little more, that is, a larger semitone. **50** From that place to the highest heaven, where there are the constellations, there is the smaller semi-

34–36 Macrob. *Somn. Scip.* 2.3.13–14.
36a Macrob. *Somn. Scip.* 2.3.15 summarized.
36b This is a reference to Porph. *Harm.*
37 Boet. *De inst. mus.* 1.1.185.4–9. Another reference to the stades is given in Pliny *HN* 2.112–113; 2.19.83 (cf. Haar, "Musica mundana," p. 120).
38 Cens. *DN* 13.2; Cap. *De nup.* 6.209.12–13.
39–40 Cens. *DN* 13.2.
41 Cens. *DN* 13.3. Cap. *De nup.* includes copious references to the stades (cf. James Willis, ed., *Martianus Capella* (Leipzig: B. G. Teubner, 1983), p. 438). Only one of these references, however, pertains to the measuring of musical intervals (cf. 2.54.20).
42–51 Cens. *DN* 13.3–4.

tone. **51** And so, from the highest heaven to the sun, the distance is the diatessaron, consisting, that is, of two tones and a semitone. **52** On the other hand, from the top of the earth to the same heaven, **52a** Pythagoreans conclude there are five tones plus two smaller semitones, from which the harmony of the diapason is made, and, moreover, Pliny makes mention of this same Pythagorean measurement in the second book of his *Natural History*.

53 Giorgio Anselmi understood these celestial intervals by means of another measurement, for he established the musical harmony of three diatonic diapasons in the motions and distances of the other planets, which intervene from the sphere of Saturn to the moon. **54** [Many aspects] in addition to what musicians practice Pythagoras related to the stars. **55** Plato and prior to him the Pythagoreans proposed that this entire universe is made through musical ratio **55a** and also by the sounds blending harmoniously together, and it consists in the agreement of sounds, as I have taken from the tenth [book] of Strabo. **56** Hence, I believe Dorilaus said that the universe is an instrument of God, **56a** when he saw that the seven planets between heaven and earth which influence the bodies of mortal men comprehend a eurhythmic motion and agreeable intervals of music and produce various sounds in proportion to each proper altitude, so concordant that they harmonize the sweetest melody, though inaudible to us because of the magnitude of the sound, which the narrowness of our ear can by no means grasp.

Chapter **Three**
Musica humana

1 The Pythagoreans and the Platonists handed down that the proper conjunction of the immortal soul and the human body and its elements is accomplished by a musical ratio. **2** Every single person looking into himself comprehends this. **3** Nothing blends that incorporeal vivaciousness of nature with the body, unless it is a certain accurate joining-together, as, for example, the tempering of low and high pitches that produces one consonance. **4** Nothing else binds together the parts of the soul itself [b1r], **4a** which, as

52 Cens. *DN* 13.4; Pliny *HN* 2.20.84.
52a Cens. *DN* 13.4.
53 Ans. *De mus.* 1.153.
54–55a Strab. *Geogr.* 10.3.10.
56 Cens. *DN* 13.5.
56a Cens. *DN* 13.1.
2–5 Boet. *De inst. mus.* 1.2.188.26–30; 189.1–5.
4a Boet. is presumably referring to Arist. *De an.* 432A.30. The reference is definitely not to Arist. *De an.* 407B–408B, Arist.'s refutation of the notion that the soul is a harmony.

Aristotle concludes, is composed of the rational and the irrational. **5** Likewise, nothing else mixes together the elements of the body or holds together its parts with one another by a certain congruity. **6** Just as a consonance in sounds is a mixture of high and low pitches pleasantly and uniformly entering the ears, as it is stated in the first [book] of Boethius's *On Music,* so the accurate joining-together of parts or the [sonorous] agreement of every single [component] is said to be a consonance, **6a** and this also happens as much in successive as in permanent matters, **7** for when two objects [of different sizes], such as wood, stones, and whatever else made from nature are put together among themselves by an artificial structure, this connection and adaptation is called by Themistius "harmonia." **8** Therefore, when we receive that which is agreeably and suitably combined in sounds, as it has been mixed and suitably adapted to us, we are delighted with it, recognizing that we ourselves are composed in its likeness. **9** If indeed our human soul, which is never in repose, shudders at those motions which are in disagreement with our own body, it is calmed down by harmonies, for it reacts with displeasure on hearing thunder and on seeing meteors and lightning.

10 If the small string of a tetrachord is stretched out more than what is just, will not the sound in its entirety altogether be dissonant and will it not be rejected by the ears as awkward? **11** In like manner, if one makes the hole of a pipe somewhat more open or more closed than what is just, will not the melody of the pipe be altogether lacking? **12** Actually, there was so much esteem for music among the Greeks that they thought it possessed a mind. **13** Indeed, it has been touched on before how frequently music repressed anger, how many miracles music achieved against the illnesses of the body and in the affection of the soul. **14** Thus our soul, whose connection in all things is musical, has a natural relation with harmony. **15** For Plato and Aristoxenus have built the soul out of those numbers from which all harmony is made. **16** Indeed, souls are joined to bodies in a certain established ratio of numbers, as Macrobius asserted in the first [chapter] of [his commentary on Cicero's] *Dream of Scipio.*

17 As long as [numbers] exist, the body persists in being animated, **18** but when [numbers] disappear, then that hidden force by which the association stood together is set free, and this is what they called fate and fatal times in life.

6 Boet. *De inst. mus.* 1.3.189.15–29, 190.1–32, 191.1–4 summarized.
7 Barb., trans. *Them. Paraphr.* (*De an.*) 2.
8 Boet. *De inst. mus.* 1.1.180.5–9.
12–13 Boet. *De inst. mus.* 1.1.184.5–9.
14 Boet. *De inst. mus.* 1.1.186.3–4.
15 Pl. *Leg.* 7.803.
16–22 Macrob. *Somn. Scip.* 1.13.12.

19 Actually, the soul itself does not cease, since it is immortal and perpetual. **20** Yet the body grows weak when the numbers are complete, **21** and the soul is not rendered faint by animating the body, but the body will give up its duty when it can no longer be animated. **22** Hence there is that phrase of the most learned poet [Virgil]: **23** "I shall fill up the number and be returned to the shadows." **24** So, as the soul itself is joined to the body by numbers, in the same manner we have no doubt that in musical harmonies, which are confirmed to result from these numbers in a certain natural disposition and in the proper accurate joining-together, not only is the soul adapted to the body but truly the elements and also the motions are mixed. **25** Hence I believe the dictum of medical science, most distinguished in Hierophilus:

> **26** When we say something or when we are set in motion from within by the pulses of the veins, it is proved that we are joined to the virtues of harmony through musical rhythm.

27 This therefore is defined as *musica humana,* which links together the concordance of diverse elements in one composite thing, that is, the human body. **28** Through this concordance the spiritual nature is bound to the body, and the rational nature [b1v] is coupled in concord with the irrational. **29** This seems to have been believed by the Platonists, who assumed that the soul is a certain harmonic entity.

30 Pythagoras and Philolaus have written together that the soul itself is a harmony, whence it has been written down. **31** "The binding-together of body and soul is made according to consonance of the natural order, that it may stand firm."

32 For that reason, Pietro d'Abano subordinates other sciences to this natural music. **33** In fact, Socrates maintained that this *musica humana* consists solely in the voice. **34** We read that Aristoxenus, in fact confined *musica humana* to the voice and in the motion of the body. **35** Yet Theophrastus has placed this *musica humana* in the voice and also in the motion of the body and of the soul. **36** But if there is harmony in the motion of the body and the soul, it is

23 Virg. *Aen.* 6.545.
25–26 Cens. *DN* 12.4–5.
31 For *loci paralleli,* see Aristid. Quint. *De mus.* 3.24 ("That not only the body of the universe but also the soul was organized and is considered through consonant numbers, the ancient and wise men affirmed confidently"); Pl. *Ti.* 35A–C, also many Timaeus commentaries, especially in Plut. *De an. proc. in Ti.* 1013Cff. For more specific references, see the annotations in Aristid. Quint. *De mus.* 3.18–24 (cf. Mathiesen, trans., *Aristides Quintilanus: On Music,* pp. 185–199). This information was kindly communicated to the translator by Professor Mathiesen on 4 February 1992.
32 D'Abano *Concil.* 3.1.
33–35 Cens. *DN* 12.1.

without doubt necessary that music itself comes together in our birth, particularly since the number of days and months which correspond to the pregnancy of women is controlled by musical ratio.

37 Pythagoras, as Censorinus recounts, said that there are generally two sorts of births. **38** The smaller one, which they call the "seven-month birth," is when the baby emerges from the womb on the 210th day after conception. **39** The larger one of the two, that is, the "ten-month birth," is when the baby is born on the 274th day. **40** Of these, the former and smaller birth is certainly contained in the number six. **41** Now that what is conceived from the seed, is, as they say, for the first six days a milky humor; **42** then, during the following eight days a bloody humor. **43** These eight [days], when they approach the first six [numbers], accomplish the first agreement of sounds in the diatessaron [and], in the third stage, nine days approach, already produce the flesh. **44** These numbers joined with the first six numbers, make the sesquialter ratio and the second agreement of sounds, that is, the diapente. **45** Then in the course of the following twelve days the body is formed. **46** Also the bringing-together of these numbers to the same six numbers yields the third agreement of sounds, that is, the diapason, derived from the duple proportion. **47** Indeed, these four numbers, that is, 6, 8, 9, 12, added together make 35.

48 Not without reason is six the basis of producing, for certainly 6 is the only number within the ten said to be perfect, since it is formed from all its aliquot parts. **49** But as the beginnings of the seed and that milky foundation of conception are loosened first of all from this number, so this beginning of the formed human being and, so to speak, the other basis of maturing, which is 35 taken six times, when it arrives at the 210th day, is brought forth mature. **50** This other birth, however, which is larger, is contained in the larger number, that is, in 7, by which the entire human life is bounded, as Solon writes, and which the Jews follow in numbering all the days, and which the ritual books of the Etruscans seem to indicate. **51** Also, Hippocrates and the other physicians disclosed nothing different concerning the illnesses of the body in that they observe only the seventh day as the day of crisis in disease. **52** Therefore, just as the origin of the other birth occurs in six days after the seed is turned into blood, so of this [birth] in seven [days]; **53** and just as in smaller births in thirty-five days the infant is formed limb by limb, so here by means of the proportions of almost forty days. **54** For this reason, people in Greece accept the fortieth day to be notable. **55** For indeed, a pregnant woman before the fortieth day does not come into a temple, and after the birth in forty days, most fetuses are heavier **56** and occasionally do not contain blood [b2r],

37–60 Cens. *DN* 11.1–8; see also Gaf. *De harm*. 4.18.

57 and babies which are liable to sickness through this time are without laughter and not out of danger. **58** For this reason, when this day goes by without complication, people are accustomed to enjoy a festive day. **59** Then these forty days multiplied by those initial seven days yield 280 days, that is, forty weeks. **60** But since the birth is delivered on the first day of this final week, six days fall away, and the 274th day is honored.

61 Truly, although both births seem to be contained in even numbers of days and Pythagoras praises the odd number, what has been asserted nevertheless does not contradict. **62** Indeed, he says that two odd numbers, that is, 209 and 273, are completed, and to their completion something is added from the following days, which, however, does not add up to a whole day. **63** We see that nature preserves its example in the period of the year as in the period of the month. **64** Since this pattern accumulates an odd number of 365 days of the year with a little over, the thirty-one days add something to the lunar month. **65** Therefore, the seven-month birth conforms with one set of numbers [pertaining to the days of the seven months], the ten-month birth with the other set of numbers pertaining to the days of the ten months. **66** Pythagoras is of the opinion that those numbers which effect any sort of mutation in each type of birth—when either the seed is turned into blood or blood into flesh, or flesh into a form of human being—have the ratio which the musical consonances have. **67** So one must not believe that music pertains minimally to our birth.

Chapter **Four**
Musica instrumentalis

1 Experience teaches us to perform instrumental music on diverse instruments and in various fashions. **2** When one emits sounds and forms consonances by natural instruments, this is called harmonic music; **3** when one emits sounds and forms consonances by artificial instruments, this is threefold: **4** one produces these consonances either by artificial musical instruments, by rhythmic instruments, or by musical and rhythmic instruments mixed together. **5** Then the one music is called musical, the other rhythmical, and the other compounded of the two kinds.

61–64 Cens. *DN* 11.11–12.
65 Cens. *DN* 11.2.
66 Cens. *DN* 9.3.
67 Cens. *DN* 12.1.
1 The topic of instrumental music is also discussed in Gaf. *De harm.* 1.1.
2–4 A different distinction between harmonic, organic, and rhythmic music is made in August. *De ord.* 2.14 (cf. *PL* 32 (1877), col. 1014).

6 Harmonic music comprises **6a** and controls the tones produced by the human voice with the aid of natural "instruments." **7** According to the Philosopher [Aristotle], these "instruments" **7a** are the lung, the throat, the palate, the tongue, the lips, and the teeth. **8** Moreover, it is called "harmonic" from "harmony," as it is conveniently produced by a melody of the voice, that is, from a human body which is composed of diverse instruments and elements. **9** This harmonic music carries forth a motion from the soul and the body; **10** from the motion it makes sound, **10a** and from the sound word and melody. **11** The hearing is occupied and distributed in the motion of the body and in the relation of sounds and in making melody, and also in the words suitable to melody, which all together produce perfect song. **12** Therefore, harmonic music is a remarkable art that governs the pitch and also contains in itself rhythm and a certain dimension of sound; **13** [this music] is the science of complete song-making [b2v], which consists of sounds, words, and rhythms.

14 Harmonic music is taken now and again in a broader sense in connection with those instruments which bring forth harmonious music and harmony in every possible way. **15** With those instruments we also learn conveniently to combine the art, hand, and string. **16** Indeed, Augustine, in the first [book] of his *On Music,* defines this **16a** to be the science of making melody well **16b** and [says] melody [*modulatio*] itself comes from "modus," for in all modes one must be concerned lest the mode be either exceeded or not fulfilled in these circumstances which are produced by some motion. **17** Or if nothing moves, we may fear lest anything is made apart from mode.

18 Melody-making, then, is not unsuitably said to be the craft of moving or, in a fixed sense, the skill that makes something move well. **19** And we cannot say anything moves well if it does not observe proper measure. **20** Again, it is necessary to understand melody in all its virtues, because nothing is made well except by moving well, **21** and if, perhaps, music resides or is contained in all things, the name of "modulatio" is all the more proper and certain in harmonious sounds. **22** Therefore, it is established that music is the [practical] knowledge of making melody and consists of sound and song. **23** Furthermore, in the meters the poets make use of, a music exists called "metrical," **23a** about which Augustine writes sufficiently at length. **24** Cer-

7 Arist. *Poet.* 20.7; see also below 2.1.75. Arist. also wrote about the voice in his *De audibilibus* (no longer considered a genuine work of Arist.); cf. Palisca, *Humanism,* pp. 143ff.
7a Arist. *Poet.* 20.7.
16–21 August. *De mus.* 1.2.3.
16a On the origin of this dictum ("musica est scientia bene modulandi") in Varro *De mus.* (lost) 7, see Christoph von Blumenröder, "Modulatio/Modulation," *Handwörterbuch* (1983), p. 1.
23–24 August. *De mus.* 2.8; 3.1 ff; 4.1 ff.

tainly through this metric of diverse feet, such as the iambic, tribrachic, dactylic, and so forth, the musicians and poets perceive the measure by provable ratio, which governs the arrangement of tempora **24a** and easily measures out these verses through a fitness expressed in sound.

25 We see that organic music of wind instruments, such as trumpets and horns, produces musical sounds by the aid of the breath. **26** On the other hand, the rhythmic music of diverse instruments is perceived to produce sounds on stringed instruments and strings by [drawing] of a bow or plucking with the fingers. **27** And there are [also] other kinds of instruments for which it is necessary to combine air and striking of fingers in order to execute harmonious music, such as pipes and pandouras, and those in which air is also brought together by skill [and] not by nature, such as the organ, for which pairs of bellows are known to supply air.

Chapter **Five**
The musician and the singer

1 There are three classes of those who profess harmonic practice. **2** One plays instruments, **3** another composes songs, **4** and the third judges the work of instruments and song. **5** The first class is of those who work with instruments and therein resides their entire art, as kitharists, organists, lutenists, **5a** and others who demonstrate the art of these instruments. **6** These musicians are far from understanding the science of music because they bring no intellectual understanding to bear on their work. **7** Innocent of all speculative knowledge, **7a** they follow the sense and what delights it, managing to commit music to memory and move their body according to it, applying a certain power of imitation. **8** Even if these musicians [b3r] are thought to accomplish much proficiently and skillfully, it is well established that they do not have scientific knowledge; **9** what they profess or exhibit they do not possess in the purity and truth of the intellect. **10** If reason indicated that there are such theatrical performers, we shall not hesitate to deny science to them. **11** We shall by no means concede to them music, which is the knowledge of making melody. **12** Truly, knowledge exists without practice and for the most part is greater [in theorists] than in those who are outstanding in practice, though the latter can by no means arrive even at the practice without any knowledge.

25 Boet. *De inst. mus.* 1.2.189.5–13.
26–27 Boet. *De inst. mus.* 1.2.189.5–13.
1–7 Boet. *De inst. mus.* 1.34.224.25–31, 225.1–3.
7a–11 August. *De mus.* 1.4.8.
12 August. *De mus.* 1.5.10.

13 Augustine, however, attributes the faster or easier motion of the fingers not to knowledge but to practice, as knowledge occupies a place only in the intellect. **14** On the other hand, (although governed by the mind), this [practice] is [a function] of the body. **15** Not only the speed and the slowness of motion but also the way the members move must be ascribed to practice rather than to knowledge; **16** for if that were different, then anyone would better make use of his hands the more learned he becomes. **17** Nevertheless, not all who play together harmoniously on instruments are without knowledge of music, but neither do all possess the knowledge itself—that which is established in the mind. **18** Moreover, in the body lies the imitative faculty, which, however, we do not believe takes hold of what is embraced by the mind. **19** The aulete, kitharist, and organist, for example, will be able to make use of theory in the mind along with the imitative faculty, that is, what is inherent in the body.

20 The second class is that of composers, who are led to song more by their own certain natural instinct than by speculation and reasoning. **21** It is well known that our Boethius separates this [class] from speculation about music.

22 The third class are those who possess the expertise for judging, so that they are truly able to grasp rhythms and melodies and also song as a whole. **23** When something has been set up entirely through reasoning and speculation, it accords properly with the art of music. **24** Consequently, he is a musician who possesses the faculty of musical speculation and reasoning **24a** and not he who possesses only the practical method of singing. **25** For every art and all science, to which the philosopher is a witness, naturally holds reason as more honorable than the craft which is exercised by the hand and labor of the craftsman. **26** Certainly, craft is corporeal and acts as if a servant; **27** reasoning, as the mistress, rules. **28** Therefore, the labor of irrationally led hands is in vain. **29** To be sure, reasoning itself does not require action, although many aspects of this science when not reduced to action are in vain. **30** Aristotle calls that reasoning which concerns action "practical" and that which consists in understanding and speculation "theoretical." **31** Therefore, in the third [book] of *On the Soul,* Aristotle states that the speculative intellect simply considers the truth but the practical fits the truth to work.

32 Accordingly, this discipline, because it is a natural science, prefers the speculative or theoretical intellect as nobler and wiser than the practical, [as

13–16 August. *De mus.* 1.4.9.
17–18 August. *De mus.* 1.4.6–7.
19–24 Boet. *De inst. mus.* 1.34.224.28–31; 225.1–11.
25–29 Boet. *De inst. mus.* 1.34.223.28–30; 224.1–7.
30–31 Barb., trans. *Them. Paraphr. (De an.)* 3.
32 Barb., trans. *Them. Paraphr. (Metaph.)* 6.

stated in the] sixth [book] of the *Metaphysics*. **33** The speculative musician, then, is he who, led by reasoning, has acquired the science of singing not by the drudgery of practice but by the power of speculation. **34** Were he to devote his effort to practice, he would add not [merely] a diverse multiplicity to this speculation but perfection, as the commentator [b3v] on the eighth [book] of the *Metaphysics* asserts. **35** This is expressed by Themistius in the second [book] of his *On the Soul,* where he says that a wise man is he who is capable of science by nature—not one who is merely capable, however, but who has already undertaken science and possesses it.

36 Of this class we call [him] a knowledgeable geometrician who has learned this science and retains it. **37** Both [the speculative and the practical intellects] potentially know, but not in the same sense, for the former [does so] a little dimly and very tenuously only because he is teachable and capable of learning, **38** whereas the other goes beyond, because he can contemplate an action and comment on it if not prevented by an external cause.

39 Apart from these, there is a third [kind], who does not know from action alone; **40** he considers and contemplates while in action and work. **41** Lacking neither theory nor practice, he is called a true musician. **42** If such a [person], having been hindered either by intention or chance, stopped his speculation or singing, he would no more cease to be a musician than a speaker who has ended a speech [ceases to be an orator], or a physician who has relinquished the practice of curing [ceases to be a physician]. **43** Indeed, when he withdraws from work and turns to contemplation, he loses nothing, according to Themistius, but he promulgates for use the science which he possesses himself. **44** Therefore, in the former there is a loss and corruption of prior qualities **45** while in the latter a certain perfection and embellishment are perceived. **46** Certainly, he preserves what he knows when he theorizes from action, **47** and he also does not destroy knowledge by theorizing but rather cultivates and completes it.

48 The "practical" man of this discipline is the singer himself, who performs and sings what the musician, dictated by reasoning, puts out. **49** Hence harmony fails if the singer tries to perform something wherein the rational ordering of the musician has been neglected and rejected.

50 Actually, the musician is to the singer **50a** as the leader is to the herald, as Guido in his didactic poem expresses in these words:

33–34 Barb., trans. *Them. Paraphr. (Metaph.)* 8.
35–40 Barb., trans. *Them. Paraphr. (De an.)* 2.20.55–56.
43–47 Barb., trans. *Them. Paraphr. (De an.)* 2.20.57.
50 March. *Lucid.* 16.1.7; Gall. *Rit. can.* 344b (cf. Erich Reimer, "Musicus–cantor," *Handwörterbuch* (1978), pp. 5–6).

51 Great is the difference between musicians and singers: **52** the latter [merely] sing, the former know what music is made of. **53** One who does what he does not know is defined to be an animal. **54** Moreover, if they praise clever tricks with resounding voices, will the nightingale or the ass be superior? **55** As these things are so, dialectic takes up its own.

56 Thus the speculative surpasses the practical, **57** for it is by far greater and more worthy to know what one does than to do what one knows. **58** Hence, we clearly see that the construction of edifices and victories in war are attributed to those by whose order and knowledge they were brought about and not to those who by their labor and service accomplished [them].

Chapter **Six**
Art contributes more to the musician than nature

1 Even if we believe that nature contributes not a little to enable men to understand music, we must nevertheless consider that instruction is valued more. **2** But the perfect musician will not succeed without both, although nature can function without instruction and instruction without nature. **3** We believe that there is something of importance in nature, for [b4r] ignorant people as well as birds are known to sing under the guidance of nature without instruction. **4** Yet we observe that the perfect and consummate musician owes more to instruction than to nature. **5** The nature of the human character does not prevent man from arriving at this [perfect knowledge of music], **6** but even if he should not reach it, those who strive for perfection will make more progress than those who, despairing of arriving at their desired goal, come to a stop right away at the lowest level. **7** Yet rules and skills are of little value without help from nature, **7a** which, as the Philosopher says, art strives to imitate as much as possible.

8 Many things become cheapened through singing and dancing. **9** If we were to confer on them the term "modulatio," the discipline itself would become worthless. **10** For melody concerns every singer, not only the one who is a musician; **11** because when the unlearned singer harmonizes with the high and low pitches of other voices and of instrumental sounds at appropriate times, he certainly produces this according to number and with variety. **12** But if questioned about these numbers or about the intervals of high and low sounds, he is not able to answer. **13** For he sings well guided by a certain

51–55 Guido *Reg.*
58 Boet. *De inst. mus.* 1.34.224.20–25.
7 Arist. *Poet.* 20.7; see also below 2.1.75.

natural sense. **14** Likewise the nightingale **14a** and the goldfinch **14b** and many other birds, **14c** whose song is harmonious and very sweet and—unless I am mistaken—rhythmic, make good melody; **15** yet they do not possess the experience in this liberal art, because we see that they simply follow nature.

16 It is not legitimate to compare those who sing accompanied by the aulos, kithara, or such instruments to the nightingale, **16a** as they have a certain art in them, for, as was said, in the nightingale there is only nature. **17** But even if often a number of people take great pleasure in the music of these birds who are altogether inexpert in the art, and this song is quite arbitrary and adheres to no law, and no regulated difference of pitches is reduced to harmony, still it seems to please them no less. **18** This is no wonder, for we observe that certain brute animals do the same; **19** indeed, they are moved to song, for example, elephants, bears, and many other animals of this kind; **19a** even cranes are charmed with great delight by their voices.

20 If nature is joined and cemented to art, will not the songs be pleasanter, sweeter, and more acceptable to our ears by far? **21** What is more, they will not be considered human but divine, and they will display those sounds of the heavenly orbits that the Pythagoreans proclaim so much. **22** So it is of very great importance whether the singer, whoever he be, is a musician or not. **23** Indeed, I dare to affirm that a musician is able to rise above nature itself in making melody and in improving these sounds through study, diligence, and industry, as we read of what Demosthenes did. **24** By the same token, no one should think it ridiculous if all very learned men have unanimously defended in public that art contributes much more to the musician than nature does. **25** This [fact] was [even] agreeable to Fabius [Quintilian] where he proves that music can in no way exist without art.

Chapter **Seven**

The judgment of music

1 Every exercise of musical art is appropriate to the sense of hearing, because if it were absent altogether, a minimum of care would have been spent in dealing with the voice by philosophers. **2** Yet not the hearing alone, which is often inconsistent and variable, but the reason arrogates to itself firm

14 August. *De mus.* 1.4.5; Pliny *HN* 10.29.81; see also above 1.1.70.
14b August. *De mus.* 1.4.5. This passage is also found in *Theorica* 1.1.70.
14c, 15–16a August. *De mus.* 1.4.5–6.
18–19 August. *De mus.* 1.4.5.
19 See above 1.1.68a.
23 Quint. *Ist.* 11.3.54; see also above 1.1.275.
25 Quint. *Ist.* 1.10.

judgment. **3** Still, hearing is permitted to possess a modicum of judgment in a certain manner, so to speak, in the sense of admonition; **4** nevertheless, the ultimate perfection of knowledge depends upon reason, which, constrained by specific rules, will ever proceed without error.

5 But our Boethius believes that one must not talk too long about the error of the senses, since all do not have the same power of sensing, nor is it always equal in the same person. **6** What we truly desire to know, we eagerly seek in vain with varying judgment. **7** Thus the Pythagoreans employ a certain middle path: **8** they do not assign all judgment to the ears but grant that certain [things] be explored only by the ears.

9 To be sure, we perceive the consonance—by which intervals differ among themselves—by ear. **10** We do not measure sounds by ear, whose judgment is blunt, but by rules and reasoning, just as if the sense were some obedient servant and reason the judge and the commander. **11** Although nearly every step of science and also of life itself is produced instant by instant through the senses, yet in them there is no certain judgment and no comprehension of truth if the judging power of the intellect is absent. **12** Thus what the philosophers concluded turns out to be valid. **13** Surely sense itself, whether the object is very large or very small, can be misled. **14** For the smallest—because of the smallness of the perceptible—cannot be sensed, since it is often mixed with the larger. **15** The hearing becomes deaf with intensive noise **15a** and grasps the smaller sound with greater difficulty.

Chapter **Eight**

The investigation and discovery of musical consonances

1 The records of the Greeks agree that Pythagoras, when he examined the musical consonances, turned to the importance of rules, having excluded the judgment of the ears, which are partly altered by nature, partly by extraneous circumstances, and partly changed through the periods of life. **2** In fact, diverse instruments offered nothing trustworthy to him, because much variety and inconsistency frequently arise in them.

3 Now, if we observe the strings [of a musical instrument], either more humid air attenuates, or drier air excites the vibrations, or the magnitude of a

3–11 Boet. *De inst. mus.* 1.9.195.19–27; 196.1–10.
13–15 Boet. *De inst. mus.* 1.9.196.10–15.
1 Boet. *De inst. mus.* 1.10.196.18–27; Nicom. *Man.* 6.2; see also Gaf. *De harm.* 1.3. A more detailed discussion concerning the transmission of this important account in more recent sources is provided in: Münxelhaus, *Pythagoras musicus,* pp. 39–40.
2–3 Boet. *De inst. mus.* 1.10.196.18–27; Nicom. *Man.* 6.2.

string renders the sound lower, or a thinner string raises the pitch, or something else completely changes the steady state in a certain way.

4 Pythagoras, confused by such great variation, happened to pass a blacksmith's shop, where the sound of hammers rang in his ears. **5** Since it was his practice to investigate the nature of everything visible, invisible, and sensible, he asked the smithies to exchange the hammers. **6** When he noticed that the sounds of the hammers did not vary with the strength of the men but were produced by the different weights of the hammers [b5r], he began to examine the weights themselves. **7** He rejected the fifth of the five weights, as Boethius asserts, **7a** because it was unconsonant with all. **8** But he observed that of the four remaining hammers, that most different in weight corresponded to the duple ratio, producing the consonance of the diapason, **9** so that if the first hammer were six pounds, the last would weigh twelve pounds. **10** The second to the first hammer showed the epitrite [proportion], forming the consonance of the diatessaron, **10a** and this measured eight pounds. **11** The third to the first hammer brought forward the hemiolic proportion or sesquialter in the number 9, sounding the consonance of the diapente. **12** The same third to the second hammer showed the sesquioctaval, constituting the tone. **13** To be sure, the fourth hammer related to the third in the epitrite number, complying with the consonance of the diatessaron. **14** The fourth to the second, producing the sesquialter ratio, yielded the consonance of the diapente. **15** Thus the weights of the hammers are arranged in this order: 6, 8, 9, 12. **16** Truly, on account of the intense thought on such important and hidden matters, the long investigation had tired even Pythagoras, I am easily convinced that he had investigated not only these minglings of the four sounds but also those remaining through which the musical discipline is perfected and maintained.

17 There are six principal relations of pitches which are derived from the multiple and superparticular proportions. **18** Macrobius (in his commentary on the *Dream of Scipio),* Chalcidius (in the interpretation of Plato's *Timaeus*),

4 Boet. *De inst. mus.* 1.10.197.3–12 paraphrased; Nicom. *Man.* 6.3.
5 Boet. *De inst. mus.* 1.10.197.3–12 paraphrased.
6 Boet. *De inst. mus.* 1.10.197.3–12 paraphrased; Nicom. *Man.* 6.4.
7 Boet. *De inst. mus.* 1.10.197.21–22; Nicom. *Man.* 6.3.
7a Boet. *De inst. mus.* 1.10.197.22.
8–9 Boet. *De inst. mus.* 1.10.197.13–14; Nicom. *Man.* 6.6.
10, 11 Boet. *De inst. mus.* 1.10.198.4–6; Nicom. *Man.* 6.7, 6.9.
12 Boet. *De inst. mus.* 1.10.198.6–8; Nicom. *Man.* 6.8.
13 Boet. *De inst. mus.* 1.10.198.3–5; Nicom. *Man.* 6.7, 6.9.
15 Boet. *De inst. mus.* 1.10.197.27–29.
18–19 Macrob. *Somn. Scip.* 2.1.9–14; Chalcid. *in Tim.* 45 (cf. Münxelhaus, *Pythagoras musicus,* p. 45).

and all musicians venerate these proportions with the greatest eagerness. **19** They concede that these proportions are easily derived and must be drawn together one by one in succession: the diapason, the diapason-plus-diapente, and the bisdiapason; the diatessaron, the diapente, and the tone from the duple, triple, quadruple, epitrite, hemiolic, and epogdoic, respectively. **20** Therefore, we may suppose that Pythagoras himself entirely changed the inconsonant fifth hammer into another and added a sixth above this fifth, which has been established in the number 4, which would surpass all others in smallness of weight, and also a sixth hammer, which would exceed all the others, that is, in the number 16.

21 Thus the progression of these six hammers occurs in their proper weights in this order, namely, 4, 6, 8, 9, 12, 16. **22** To be sure, if Pythagoras juxtaposed the eight-pound hammer and the four-pound hammer, he enjoyed hearing the consonance of the diapason in this duple ratio, **23** and when he arranged the sixth hammer [weighing] sixteen pounds with the eight-pound hammer, he similarly achieved the diapason in this duple proportion. **24** When he compared this heavier hammer to the lighter one, that is, of four pounds, he discovered the bisdiapason in the quadruple ratio. **25** But if he coupled the twelve-pound hammer to the first of four pounds, he brought forward the consonance of the diapason-plus-diapente in the triple proportion.

26 In this manner, he found out that when these different weights were laid out, all the musical consonances were contained solely in the multiple and in the superparticular genus; **27** and the heavier hammers sounded against the higher in a certain mutually corresponding order, so that both returned to the sense of hearing with an established dimension [b5v] by presenting out of the different blows a single consonant sound.

28 When these facts had been confirmed as true, Pythagoras investigated whether the ratio of all consonances consisted in these proportions. **29** He fitted to equal strings weights arranged according to the same numbers and proportions. **30** He gained perfect confirmation, then, in a separate experiment, fitting the length of several reed pipes to each ratio. **31** He repeatedly measured vessels of equal weight, proportionally filled with water, which he hit with a rod. **32** He also hit brass bells of different weights with a brass and iron rod and noticed that musical consonances of the same type are joined in these proportions **32a** and respond harmoniously. **33** Hence the conclusion

29 Boet. *De inst. mus.* 1.11.198.13–21 paraphrased; Nicom. *Man.* 6.5.
30 Boet. *De inst. mus.* 1.11.198.13–21 paraphrased; Nicom. *Man.* 6.11.
31 Boet. *De inst. mus.* 1.11.198.13–21 paraphrased.
33 Boet. *De inst. mus.* 1.11.198.21–22.

Figure 1.8.1

derived is more certain in that the length and the thickness of brass and gut strings, regulated by these proportions, produce musical sounds.

34 In this way, Pythagoras discovered the monochord rule [*regula*] which gets its name "rule" from the object not because a rule is made of wood or brass and we measure sounds and magnitudes with it but because a rule is a certain fixed and stable observation that leaves the judgment in no doubt. **35** It is so named from "ruling" [*regendo*], as if it ruled us in such a way that we cannot fall into error. **36** However, Josephus attributed this Pythagorean investigation to Jubal before the Flood. **37** Surely, Jubal himself feared the Flood—Adam having predicted to the human race that the universe would perish in a quick fire or in the sea—and cherished the most pleasing and extraordinary art of sounds to such a degree that he engraved these consonances and proportions on two marble columns and also on a brick sculpture—**38** marble so that it would not rot in the Flood, and brick so that it would not be destroyed by fire. **39** After the Flood, posterity found this sculpture again and put it to use. **40** Figure 1.8.1 shows the forms of those sought out proportions and deduced consonances.

34 Boet. *De inst. mus.* 1.11.198.23–28. Boet. discusses this topic further in *De inst. mus.* 4.5–12.

36–39 Com. *Hist.* (cf. *PL* 198, col. 1079).

BOOK **TWO**

Chapter **One**
The definition of music and properties of sounds

1 Every instruction presented to the reason on any subject ought to proceed from a definition. So that what we discuss may be understood, we propose that the discipline of music, which Pythagoras and Plato called philosophy, is in the present purpose a science of numbers proportionate to sounds that measures the differences of low from high sounds through sense and reason. **2** Indeed, this science is called "musica" (I use the Latin word), derived according to certain writers from "moys." **3** Others, however, derive the word **3a** from the "Muses" **3b** or "apo ton muson," as if from "inquiring," because through this "musica" the law of song and melody have been sought. **4** Certain elements of this musical science are contained in nature, some in our usage.

5 The height and depth and the "diastemata," that is, the intervals, are considered inherent in nature, as Bacchius stated. **6** But the performance and the moral effect of the practice of these sounds depend on our usage.

7 The semitones, tones, systems, and consonances themselves are made from these sounds. **8** It is said that consonance through motion is the beginning of music, and consonance through attribution the end. **9** Surely music cannot be produced without sound, **10** and sound is not made save by a blow and percussion. **11** A blow and a percussion, in turn, cannot be present, if motion has not preceded, **12** for if everything were immovable, one thing could not meet another so that the one is hit by the other. **13** If all is at rest and without motion, it is necessary that no sound can occur.

14 For this reason, in the fifth [book] of [Aristotle's] *Physics,* motion

2 March. *Lucid.* 1.6.2.
3 "Apo ton muson," meaning "from the Muses."
3a Isid. *Etym.* 3.15.1.
4–6 Bur., trans. Bac. *Intro.* 2. This translation is no longer extant (cf. Palisca, *Humanism,* p. 112).
10–11 Gell. *NA* 5.15.7–8 (cf. below 2.4.20–21).
14–15 Arist. *Ph.* 5.4.228b. For another reference to the same passage in Arist. *Ph.,* see Apollinaris Offredi Cremonensis's *Quaestiones super libris Aristotelis De anima I–III,* a fifteenth-century document preserved in Rome, MS. Vat., lat. 2158, fol. 84 verso (cf. Michael Wittmann, *Vox atque sonus: Studien zur Rezeption der aristotelischen Schrift "De anima" und ihre Bedeutung für die Musiktheorie,* 2 vols., vol. 4 of *Musik-*

is called a successive changing which occurs in time, **15** for time accompanies motion. **16** In the fourth [book] of the *Physics,* time is [measured] as number of motion.

17 Consequently, sound, with Boethius, is a percussion of air that is continuously spread out up to the hearing. **18** Because this percussion is a matter of sensory perception **18a** and passes with time, it is necessary that it be impressed on the memory so that it does not vanish, as it cannot be written down. **19** It is usual to doubt whether the struck body or the striking one yields the sound or whether both of them do.

20 Themistius, however, in the second [book] of the *On the Soul* (in the translation of Ermolao Barbaro, a man most eminent in every branch of learning), says that sound is air struck and expanded. **21** Indeed, there are two kinds of air, as he says: **22** the one, which is struck between the bodies in conflict, is dispelled, **23** and the other, which is first struck by this air and soon carried by the succession of the motion directly to the hearing; **24** and this [second type] is then that which is heard.

25 Yet the difference of sounds can be recognized not when the sounds subside in potential and are pressed within, but when they are pushed forward in act, which applies also to other sensible things. **26** The principal [characteristics of sounds] are height and depth, **27** terms we borrow from sensible things and assign to sounds through metaphor. **28** Just as a sharp sword point attacks violently in a short time, but a dull one is slow and sluggish, [c1r] so a high pitch hits the hearing and vibrates violently and almost wounds it as if piercing it. **29** The low sound dulls and spreads out just as if it were pounding on the hearing. **30** In this way one sound moves very quickly, the other moves very slowly. **31** Therefore celerity and height of pitch are not the same, and neither are retardation and depth, as it seems to certain [theorists]; **32** but it thus happens that the motion of the more pointed and conical body is frequent and very swift; **33** and that of the blunt and retarded is lingering and slow. **34** And for this reason, it also happens that the higher sound is produced from more polished and broader bodies, as air is moved more swiftly from these bodies and bursts asunder more quickly; **35** whence the sense of hearing is excited vehemently and quickly.

---

wissenschaftliche Studien, ed. Hans Heinrich Eggebrecht (Pfaffenweiler: Centaurus-Verlagsgesellschaft, 1987), 2:198).

16 Arist. *Ph.* 4.11.219a. For another reference to the same passage in Arist. *Ph.*, see Marcus Antonius Genoa's *Aristotelis Esculapius in Aristotelis De anima speculationes III,* a lecture recorded in Rome, MS. Vat., cat. 4705, fol. 218 recto (cf. Wittmann, *Vox atque sonus* 2:296–297).

17 Boet. *De inst. mus.* 1.3.189.22–23; Nicom. *Man.* 4.1.

20–24 Barb., trans. *Them. Paraphr. (De an.)* 2.29.85.

25–35 Barb., trans. *Them. Paraphr. (De an.)* 2.30.86.

36 So Boethius considered that among these motions, some are more rapid, others more tardy; **37** and also of these motions, in consequence, some are looser, others denser. **38** When we turn our attention to motion in continuity, it is necessary that we detect either velocity or tardiness. **39** Every dispersed and slow motion yields low sound, **40** for as tardiness is very close to stillness, just so is depth close to silence. **41** This is accomplished by the depth itself and by the tardiness of striking. **42** A fast, swift, and dense motion, on the other hand, renders high sounds by means of celerity itself and by velocity of striking. **43** Thus, if we pull a taut string tighter, it will bring forth a high sound; **44** loosened, it yields a low sound. **45** Therefore the tenser string prompts the swifter pulse and is vibrated more quickly; **46** it also strikes the air more frequently and in denser succession. **47** Surely a less tense string brings forth loose and tardy pulses; [the pulses are] also rare, owing to this feebleness of striking; **48** nor does [the string] tremble very long.

49 Nor are we to believe that each time a string is struck only a single sound is emitted or that there is only one percussion, **49a** because the air is struck as often as the vibrating string strikes it. **50** Since the velocities are joined to sounds, **50a** no interruption is sensed by the ears; **51** for one sound, either low or high, strikes the sense, although either sound consists of many parts: **52** those of the low more tardy and rarer, the high swifter and denser [motions].

53 Likewise, it is worthwhile to distinguish that only the sound of a living being is properly called "voice." **54** Certainly, inanimate things cannot be "vocal," but if we say that small stringed instruments or small pipes are "vocal," we do so metaphorically and do not apply the word strictly, because words and utterings often seemed to be formed on those instruments. **55** This occurs because these small strings form harmonic sounds by means of tensions and relaxations according to a proportion true in both art and nature, which also obtains in pipes and other instruments.

56 Among the animals, also, many do not have a voice, **57** and these are chiefly bloodless and do not have the capability of emitting any voice, as is well known from insects and from those animals which are protected by a hard shell, **58** for those which seem to sing, such as the "cicadas" [tree-crickets], have a membrane sideways enclosed underneath [the shell]. **59** The air enclosed within this membrane is rubbed against it; **60** and so, the "cicadas" make a noise. **61** Fish also do not emit a voice; **62** and this is not

36–52 Boet. *De inst. mus.* 1.3.189.23–29, 190.1–15.
53–54 Barb., trans. *Them. Paraphr. (De an.)* 2.30.86.
56–64 Barb., trans. *Them. Paraphr. (De an.)* 2.31.88.
57 On bloodless animals, see Arist. *Hist. An.* 4.1–7.
58 For an earlier reference to the tree-cricket (*cicada*), see above 1.1.58. A more detailed discussion of the tree-cricket is provided in Arist. *Hist. An.* 4.9.535a,5.

surprising, since voice is the motion of air. **63** Fish do not take in air, considering that they spend their lives in liquid, **64** for those fish which the river Archelous brings forth, called "capri," excite what is credited to be their voice and grunting by the rubbing of the gills. **65** Consequently, only animals emit a voice—but not every animal (as has been said), nor [c1v] from every part of the body; **66** for even hands clapped together make a noise.

67 There are three things through which not only all sounds but also the voice are produced. **68** The first is that which inflicts a blow, second, that on which the blow is inflicted, and third, that in which that blow is put in motion. **69** It is plainly necessary that both bodies have air around them, although not in the same manner in both. **70** For not every kind of air is required in order to bring out the voice, **71** but only that breath [*spiritus*] which we alternately draw in and return in respiration. **72** The twin use of this [breath] is necessary in respiration, **73** because one is for cooling off the warmth which is within, **74** and the other, though not necessary, is suitable for forming the voice, as Themistius, thanks to his superior knowledge, reports. **75** The instrument of respiration is the gullet. **76** We breathe the spirit [*anima*] by alternately inhaling and exhaling. **77** It seems that the gullet is joined to the two lobes for the sake of the lungs, because animals that are active and supplied with blood keep this part rather hot, as the lung is very close to the heart, where the source of heat is. **78** Therefore, it is the task of the lung and even more so of the heart to ventilate. **79** Finally, respiration is necessary for the preservation of life. **80** Thus air, having been brought in and led through the gullet, renews this heat and cools it off. **81** Consequently, [there is] the percussion and striking of this air, which we take and return to the trachea in breathing; **82** if this happens deliberately and with attention, then the breath which is going in and out becomes the voice.

83 Surely a cough is not voice, however much it may come forth from the same places **83a** and organs that accomplish respiration with the trachea, which belongs to the lung, **84** and it occurs through the same mechanism, since, of course, coughing does not occur by intention. **85** But neither coughing, nor concave hands, which may be struck and make a noise, nor applause, nor any other sound accidentally and by chance elicited from the throats or lips are voices in the proper sense, **86** for the voice which is produced without knowledge or reflection is not a voice, as is universally established in the very definition of voice, because voice is sound which is uttered with some significance.

65–82 Barb., trans. *Them. Paraphr. (De an.)* 2.31.88.
75 The gullet is classified as an instrument by the Philosopher; cf. above 1.4.7.
83, 84–86, 88–89 Barb., trans. *Them. Paraphr. (De an.)* 2.31.90.

87 From Themistius, as translated by Ermolao [Barbaro], let us derive a more general definition of voice. **88** Themistius says that voice is the sound of any living being which is drawn from the instruments, paths, and formation of breathing, having its origin at the place where that which the breath has given back vehemently strikes air contained in the trachea or the trachea itself—**89** and that with the intention, that is, through some imagination or deliberation, of signifying or indicating something.

90 Accordingly, high pitches are incited by rather dense, swift motions, **91** the low, certainly by tardier and more scattered motions, **91a** as Boethius asserts. **92** It happens that low pitch is made higher through some increase in motion. **93** In fact, by decreasing the motion, a high pitch is made lower, **94** for it is known that more motions cause height of pitch rather than depth. **95** Surely, this higher pitch depends on a certain largest number, in which there is a difference of plurality. **96** In truth, all paucity is related to plurality in the same fashion as a number to another. **97** Among things that are compared according to number, however, some are equal, others unequal. **98** Therefore sounds are also partly equal and partly different by inequality; **99** but from these voices which transmit no dissimilitude to the ears, no consonance occurs at all, **99a** according to Nicomachus, for consonance is the concord of dissimilar voices reduced to unity.

Chapter **Two**
Sound without pitch and the distinction of pitches

1 It is well known that sound, of which we are speaking further, is twofold, action and potential. **2** Some materials preserve sound neither in action nor in potential, such as sponges and all sorts of wool. **3** Other materials, although

90 Boet. *De inst. mus.* 1.3.190.21–23; cf. above 2.1.42.
91 Boet. *De inst. mus.* 1.3.190.23; cf. above 2.1.39.
92 Boet. *De inst. mus.* 1.3.190.23–24. This remark follows logically from 2.1.43 above.
93 Boet. *De inst. mus.* 1.3.190.24–25; See above 2.1.44.
94 Boet. *De inst. mus.* 1.3.190.25–26. This remark follows logically from 2.1.42 above.
95–96 Boet. *De inst. mus.* 1.3.190.26–30.
97 Boet. *De inst. mus.* 1.3.190.30–31. Boet. here refers to his *De inst. arith.* 1.21 (cf. below 3.2.41 ff.).
98–99 Boet. *De inst. mus.* 1.3.190.31–32, 191.1–4.
99a This is a reference to Nicom. *Man.* 12.4: "Systems are consonant when the notes comprising them, though different in compass, are, when struck simultaneously or sounded in the same way, intermingled with one another in such a manner that there is produced from them a single sound like one voice." Nicom.'s definition is based upon one provided in Aristox. *Harm.* 1.17 (cf. Flora Rose Levin, "Nicomachus of Gerasa: Manual of Harmonics (Translation and Commentary)," Ph.D. diss., Columbia University, 1967, p. 259).
1–9, 10–16 Barb., trans. *Them. Paraphr. (De an.)* 2.26.77.

they have sound, nevertheless have it sometimes in action, sometimes in potential, such as air and stones and all flat, smooth, and polished things, which, when at rest, contain sound in potential. **4** Surely, when they are mutually struck, they emit sound and noise in action. **5** Indeed, sound is said to have been emitted when, having advanced from the sounding object to the ear, it fills the intermediate space. **6** But at least three things are necessary for sound to be made: two that strike together and a third in which this striking occurs; **7** those that strike together must be hard and solid bodies. **8** It is in air that this occurs. **9** A blow or percussion cannot be accomplished without a separation nor a separation without air in the middle, **9a** as Themistius proves. **10** Not every percussion produces sound, nor do all bodies send forth sound. **11** Examples are bodies that have been planed, are light, or are either hollow or extended in latitude **12** and whose smoothness prevents the fracture of air which is thus conveyed whole and undivided to the hearing. **13** But those bodies which are extended and by the extent of their borders and edges propel much air resonate excellently. **14** In like manner, hollow objects are suitable because the blows are repeated from the initial blow through the bursting asunder, **15** since, in fact, that portion of the air which is excited within the hollow body seeks an exit which it is not able to find because of the opposition of the sides on which it strikes as it hastens. **16** It happens that through this for a long time, the wrestling and twisting makes much noise. **16a** This principally occurs in connection with small bronze pots, especially if they are smooth and polished.

17 Beyond this, I consider the following to be worth knowing. **18** An echo is returned for this reason: **19** air, beaten by a sound or a voice, as often as it strikes against a body which is plane, smooth, and simple, cannot proceed beyond this obstacle, which prohibits passage through or splitting of; **20** the air is dispersed backward and bounces like a ball. **21** Indeed, it is necessary that voices always return similarly and reciprocally, **22** but because of their thinness they escape perception. **23** Certainly, air that has been struck always returns when it encounters an obstacle. **24** Otherwise, if voices always retained a straight line of travel, nobody could hear himself talking; **25** but in general the reflection is hidden, which also occurs with light. **26** Actually, an echo resounds from every noise, even if it does not always sound nor is always distinctly heard, **27** especially when the blow of air meets with a body with a curved and polished surface. **28** To be sure, if a flat and uniform surface of the body is struck, the air, too, as a whole is excited uniformly and without interruption. **29** Then the air goes back and forth unimpaired and complete in the same unit and strives to return. **30** If this body is jutting, rough, and uneven, then it is inevitable that the air splits and is torn to pieces; **31** that is, it

18–31 Barb., trans. *Them. Paraphr. (De an.)* 2.27.80.

dashes against not upon one flat surface but upon several, especially because air is [c2v] by nature most easily separable and cleavable so that it surrounds everything and fills up all space.

32 But air, as was said, is not the only thing that makes sound; **33** yet among all elements, it maintains the first place. **34** For this reason, those who assign the characteristic and special property of hearing to a vacuum are not far from the truth, because they believe that vacuum is nothing but air. **35** Therefore, the property of sounds and noises is that which can uniformly and continuously move air all the way to the ear. **36** Thus, the nature of hearing is bounded to and related to air, **37** which is held in the ears, mixed in from birth, woven as if in a breadbasket and imbued with that spirit. **38** Also air is built into the organs of sensation, where the original power of sensation is located. **39** Accordingly, as often as the exterior air is excited and penetrates into that interior, it becomes continuous and joins with the inner air in such a way that both become one. **40** Truly, just as water to water becomes continuous but not contiguous, so also air to air. **41** Consequently, the outer air transmits the sound through the inner air to the sense. **42** Also this is the reason why an animal does not hear with every member but only in that place into which nature has woven a portion of air. **43** Indeed, air itself is without sound when it is cut off from the place where sound originates.

44 Under the circumstances, the following, it seems, must be discussed: **45** when we recite something, for example, this verse, **46** "Let us, Christ, be aided by the merits of Father Ambrose," **47** where now are the meters and feet? **48** In the sound which was heard? **49** Or rather in the sense of the hearing which extends only as far as the ears? **50** Or also in the act of pronouncing? **51** Or because it is recognized as verse? **52** We must acknowledge that the numbers of rhythms exist in our memory also. **53** It may happen that a sound of this kind emerges from anywhere in little delays and measures, vibrating the air by either falling drops or by some other impulse of bodies where no listener is present.

54 There is yet another [sound] in the sense of hearing: **55** since the ears have the power of perceiving if anything sounds—which they do not lose if sound is absent—nevertheless, when there is silence, they are in some way different from deaf ears. **56** If nothing sounds, the hearing will nevertheless possess rhythms, although melodious sound is perceived. **57** If you touch a

32–43 Barb., trans. *Them. Paraphr. (De an.)* 2.28.81–83.
46 "Da patris Ambrosii meritis nos Christe iuvari." This hexameter resists closer identification (cf. Hans Walther, *Lateinische Sprichwörter und Sentenzen des Mittelalters in alphabetischer Anordnung* (Part I: A–E), vol. 2 of *Carmina Medii Aevi Posterioris latina* (Göttingen: Vandenhoeck & Ruprecht, 1963), p. 589). This information was kindly communicated to the translator by Dr. Heinz Ristory on 19 December 1991.
47–53 August. *De mus.* 6.2.2. August.'s "actu" has been replaced by "ictu" (cf. 2.2.50).
54–66 August. *De mus.* 6.2.3.

sensible spot of the body with the finger a certain number of times, it is sensed that number of times; **58** and if it is sensed there, the perceiver does not lack separation there. **59** But is not this sense inherent, even when no one touches? **60** And similarly rhythm?

61 In the sixth [book] of his *On Music,* Augustine stated that it is not easy for the perception to lack such rhythms which have not been established in it, even before anything sounds. **62** Certainly, in no other manner is hearing either pleased by continuity of sound or offended by cacophony. **63** Accordingly, for this very reason we either approve or disagree not by reasoning but by nature. **64** When anything sounds, we call this rhythm a property of the sense itself. **65** This power of approving and disapproving when I hear a sound, therefore, does not reside in my ears. **66** Surely, the ears are open to good sounds just as to bad ones. **67** Certainly hearing differs from not hearing, just as hearing this voice differs from hearing another. **68** Consequently, this disposition is neither extended beyond nor held within because it is the measure of sound which produces it. **69** Indeed, one measure occurs in the more lengthened sound, another in the shortened one, [c3r] and no measure in silence. **70** Truly, if a sound is rhythmic, it is necessary that the disposition for sensing sound be rhythmic also. **71** And this cannot exist without the producer of this sound. **72** But that natural and inherent power which belongs to the ears is not absent in silence, **72a** as has been touched on beforehand in [Chapter] five of [Book] one; **73** nor has sound planted this in us, **74** but sound must afterwards either be approved or rejected by this natural and inherent power. **75** Hence, it must be acknowledged that those elements which are in that particular perception of the ears are the rhythms; **76** when anything is heard, they are introduced by sound and removed by silence. **77** Therefore, those rhythms which reside in the sound itself can exist without those that are in the act of hearing, but the latter by no means can exist without the former. **78** But there is a third genus to be considered, which exists in the use and process of pronunciation, and also it must be seen whether these rhythms can exist apart from those which are retained in the memory. **79** Certainly, we can in ourselves silently form certain rhythms by thinking in that interval of time in which they would be formed by the voice. **80** These rhythms obviously exist in a certain operation of the soul, which, because it produces no sound at all and plants no sensation in the ears, shows that this genus can be without the other two, **81** one of which is in sound, the other in the hearer at the time of hearing. **82** Yet we ask whether one or the

67–72 August. *De mus.* 6.2.3 abbreviated.
72a Cf. above 1.5.18.
73–77 August. *De mus.* 6.2.3.
78–99 August. *De mus.* 6.3.4.

other rhythm would not exist if memory did not help. **83** If the soul forms those rhythms which we discover in the beating of the veins, [namely the blood pumping through the veins], the question is solved. **84** For it is apparent that these rhythms exist in the operation of the soul, **85** and we are helped in no way by the memory to form them. **86** If it is uncertain whether these rhythms belong to the operating soul, no one doubts that those we produce with a certain alternating motion of the breath are rhythms in intervals of time, **87** and the soul produces them. **87a** Thus when the will is brought into action, they can be varied in many ways. **88** Yet memory is not needed to produce them. **89** It seems that this genus can do without the other three. **90** Although I would not doubt that various pulses of the veins and intervals of respiration occur in proportion to a mixing of meters, **91** still, I do not dare to deny that this is accomplished by the soul in motion. **92** Also, this pulse, in view of the diversity of bodies, is faster for some persons and slower for others. **93** Unless there is a soul at hand to produce this pulse, there is none.

94 Yet a fourth genus ought to be considered—those rhythms that exist in memory. **95** Obviously, if we bring these forth by recollection and then turn to other thoughts, leaving aside those almost hidden in their secret recesses, **96** I do not believe that these rhythms can be hidden without the remaining rhythms, **97** nor can there be a doubt that they can exist without the remaining rhythms. **98** Unless they are heard or thought, however, they are not committed to memory. **99** Therefore, although rhythms adhere to the memory after those heard, the former were impressed on the memory by the latter.

100 It seems proper that yet a fifth genus, distinguished from the other four, be added. **101** For it is one thing to make a sound, which is attributed to the body, **102** and another to hear, which the soul experiences in the body from the sounds. **103** It is one thing to produce rhythms, either longer or shorter, **104** another to remember them, **105** and yet another to form an opinion concerning these by approving or rejecting, as if by a certain natural law. **106** Among others, this fifth genus is more worthy than the other four genera and more excellent, **107** for it cannot judge concerning those genera unless it surpasses them, **108** something that is expressed more clearly in the fifth [chapter] of [Book] one [c4r].

109 Now it is proper to approach the classification of voices. **110** Of these

100–105 August. *De mus.* 6.4.5 paraphrased.
106–107 August. *De mus.* 6.4.6 paraphrased.
108 Cf. above 1.5.22 ff.
109 Boet. *De inst. mus.* 1.12.199.1–2; The following classification of the motion of the voice according to one of three species, namely, continuous, diastematic and intermediate, is due to *Aristox. Harm.* 1.8.10, 18–19, but is also transmitted in several later sources (cf. Mathiesen, trans., *Aristides Quintilianus: On Music,* p. 76.)
110 Boet. *De inst. mus.* 1.12.199.3–4; Nicom. *Man.* 2.1. Gaf. describes this motion of the voice by the term *pedestris* in *Pract.* 3.1 and in *De harm.* 1.2.

some are continuous, **111** others are spaced apart by intervals. **112** Those voices which we produce continuously in speaking or reading are continuous. **113** Of course, in reaching the senses and in expressing by talking, the impetus of the continuous voice produces its effect, hastening the words to run as fast as possible. **114** When we sing, however, we produce voices that are interrupted by an interval, through which we perform not speech but melody. **115** Indeed, pitch of this kind is slower, creating a certain interval by means of modulating varieties, not an interval of silence but of suspended and even rather slow song, as Boethius asserts. **116** To this, Albinus adds a third distinction which includes intermediate motion of the voice, **117** as those with which we recite heroic poetry, **118** by means of neither a continuous inflection, as with prose, **119** nor a suspended uttering of pitch, as with song.

120 Continuous pitches and those by which we run through a song are infinite by nature. **121** For it is evident that there is no limit either to the evolution of speech or to the elevation of its height and of the relaxation of its depth. **122** In fact, human nature has put an appropriate limit to both. **123** For the human breath sets a limit to continuous pitches beyond which they cannot be stretched out at any tension. **124** An individual speaks without interruption only so long as the natural breath permits. **125** The nature of men, likewise, has caused a limit to the sustained sounds interrupted by an interval **126** which determines the high and low pitches. **127** Actually, an individual has the capacity either to raise up the height or press down the depth only as far as the natural limit of his voice permits.

128 We read that Ptolemy has reduced the differences of successive pitches to these: **129** pitches in relation to each other are either unisons or non-unisons. **130** Of the non-unison pitches, some are equal sounding, others are consonant, some are emmelic, others dissonant, others ekmelic. **131** Unison

111 Boet. *De inst. mus.* 1.12.199.4; Nicom. *Man.* 2.1. Gaf. describes this motion of the voice by the term *equestris* in *Pract.* 3.1.
112–114 Boet. *De inst. mus.* 1.12.199.4–9; Nicom. *Man.* 2.3.
115 Boet. *De inst. mus.* 1.12.199.12–14.
116–119 Boet. *De inst. mus.* 1.12.199.14–18; Albinus, lost treatise on music (cf. Cassiod. *De mus.* 6; as cited in *PL* 70.1212). A discussion of this third type of voice is provided in Cap. *De nup.* 9.937. Gaf. describes this motion of the voice by the term *modulata* in *Pract.* 3.1.
120–127 Boet. *De inst. mus.* 1.13.199.20–27, 200.1–5. Boet. here provides a summation of Nicom. *Man.* 2.4–6.
128 Ptol. *Harm.* 1.7.15.
129 Boet. *De inst. mus.* 5.11.361.3–4; Ptol. *Harm.* 1.7.15.
130 Boet. *De inst. mus.* 5.11.361.4–6; Ptol. *Harm.* 1.4.5; 1.7.15. Concerning Boet.'s conflation of the respective passages from Ptol. *Harm.*, see Bower, trans, *Boethius: Fundamentals,* p. 170.
131 Boet. *De inst. mus.* 5.11.361.6–7. Concerning Ptol.'s definition of unison pitches in *Harm.* 1.4.5, which is distinctly different from the one provided by Boet., see Bower, trans.,

pitches are those which, having been struck, render one and the same pitch. **132** Equal sounding are those which, having been struck at the same time, yield somehow one simple sound from two, as, for example, the diapason and the bisdiapason. **133** Consonant pitches are those which yield a composite, mixed, and pleasant sound, **133a** for example, the diapente and the diapason-plus-diapente. **134** Emmelic are those which, although not consonant, are nevertheless well suited to song, **134a** for instance those which join the consonances. **135** Dissonant are those whose sounds do not mix and wound the sense unpleasantly. **136** Ekmelic are those which are not admitted in the conjunction of the consonances.

137 Consequently, there are unison pitches whose sound is one, whether low or high. **138** Non-unison are those pitches of which one is deeper and the other higher. **139** The difference between some of these pitches is joined by means of a common border, which, however, is not discrete or interrupted by an interval, **140** but it is led from the low to the high in a way that seems continuous. **141** Indeed, the pitches are joined in this manner to a common border. **142** If someone strikes the string on a musical instrument and tightens it at the same time, the sound of the string, initially lower, rises as it is tuned up. **143** The sounds of the low pitch and of the high then become continuous. **144** An analogous image is set before the eyes **145** when [c4r] we observe a rainbow in the clouds; thus, the colors are next to each other so that there is no definite border by which one color is separated from the other; **146** rather they

---

Boethius: Fundamentals, pp. 170–171. In *Pract.* 3.1 Gaf. refers to Bryen.'s classification of this motion of voice as *homophoni* or *antiphoni*.

132 Boet. *De inst. mus.* 5.11.361.8–10; Ptol. *Harm.* 1.7.15. Concerning Boet.'s phrase "pulsae . . . simul," not found in Ptol.'s treatise, see Bower, trans., *Boethius: Fundamentals,* pp. 47, 116, 171, 173. Further on this topic, see below 5.4.50–61, as well as Gaf. *Pract.* 3.1..

133 Boet. *De inst. mus.* 5.11.361.10–12; Ptol. *Harm.* 1.4.10 paraphrased. In Pract. 3.1 Gaf. refers to Bryen.'s classification of this motion of the voice as *symphonia*.

133a Boet. *De inst. mus.* 5.11.361.12; Ptol. *Harm.* 1.7.15. In both *Theorica* and *Theoricum* Gaf. has replaced Boet.'s "diatessaron" by "diapasondiapente."

134 Boet. *De inst. mus.* 5.11.361.12–14; Ptol. *Harm.* 1.7.10. In *Pract.* 3.1 Gaf. refers to Bryen.'s classification of this motion of the voice as *emmeles* or *concinni*.

134a Boet. *De inst. mus.* 5.11.361.14–15; Ptol. *Harm.* 1.7.15.

135 Boet. *De inst. mus.* 5.11.361.15–16; Ptol. *Harm.* 1.4.10.

136 Boet. *De inst. mus.* 5.11.361.16–17; Ptol. *Harm.* 1.4.10. Concerning Ptol.'s definition, which differs from Boet.'s wording, see Bower, trans., *Boethius: Fundamentals,* p. 171.

137–140 Boet. *De inst. mus.* 5.5.356.7–12; Ptol. *Harm.* 1.4.9.29–10.12D (cf. Wille, *Musica Romana,* p. 693); see also Gaf. *De harm.* 1.2.

141 Boet. *De inst. mus.* 5.5.356.13–14; Ptol. *Harm.* 1.4.9.8fD (cf. Wille, *Musica Romana,* p. 693).

142–143 Boet. *De inst. mus.* 5.5.356.20–24; Ptol. *Harm.* 1.4.9.8fD (cf. Wille, *Musica Romana,* p. 693).

145–146 Boet. *De inst. mus.* 5.5.356.14–20; Ptol. *Harm.* 1.4.9–10. Concerning Boet.'s deletion of Ptol.'s further remarks on low and high sounds of the continuous type, see Bower, trans., *Boethius: Fundamentals,* p. 167.

go from red to yellow, for example, so that one color turns into the next in uninterrupted mutation, without any definite intermediate color that might distinguish both.

147 The difference between other non-unison pitches is distinguished through an intervening silence. **148** Therefore, whereas some pitches are continuous, others are discrete. **149** The continuous are of such a nature that the difference between them is joined in a common border. **150** But discrete pitches occupy their proper locations in relation to each other, like unmixed colors, whose differences are observed to lie in some nonarbitrary relation. **151** Thus these pitches differ from each other by their height and depth **152** and are separated by an interval in between. **153** In them depth is created from height and height from depth. **154** Thus high pitch and low do not occupy a special and designated location. **155** Continuous and non-unison pitches, however, cannot correspond to the harmonic property at all **156** when they are dissimilar to each other and do not sound in any single [pitch]. **157** Discrete pitches are subject to musical art, however, **158** for the difference of unlike pitches, which are separated and dissimilar, can be perceived.

Chapter **Three**
Lowness, height, and interval

1 They say that the tone, which the Greeks call *phthongus,* from their corresponding word for "speaking," is the inflection of emmelic or melodious pitch, falling upon one tension. **2** One such tension, achieved on a given pitch, yields the emmelic phthongus, that is, one that is harmonious and melodious, as Bacchius says. **3** Therefore, when we consider consonance to be a mixture of high and low tones sweetly and also uniformly reaching the ear, **4** a certain

147 Boet. *De inst. mus.* 5.5.356.12–13; Ptol. *Harm.* 1.4.9.29–10.12D (cf. Wille, *Musica Romana,* p. 693).
148–149 Boet. *De inst. mus.* 5.6.356.26–27, 357.1; Ptol. *Harm.* 1.4.9–10.14D (cf. Wille, *Musica Romana,* p. 693).
150 Boet. *De inst. mus.* 5.6.357.3–5; Ptol. *Harm.* 1.4.10.14D (cf. Wille, *Musica Romana,* p. 693); see also Gaf. *De harm.* 1.2.
154 Boet. *De inst. mus.* 5.6.357.1–2. Boet. refers to Ptol. *Harm.* 1.4.10.13D (cf. Wille, *Musica Romana,* p. 693).
155 Boet. *De inst. mus.* 5.6.357.5–6; Ptol. *Harm.* 1.4.10.23 (cf. Wille, *Musica Romana,* 693). Boet.'s erroneous "non aequisonae" (and Gaf.'s adoption of Boet.'s error) requires an emendation to "non unisonae."
156–158 Boet. *De inst. mus.* 5.6.357.6–9; Ptol. *Harm.* 1.4.10.23D (cf. Wille, *Musica Romana,* p. 693).
1 Boet. *De inst. mus.* 1.8.195.2–5; Nicom. *Man.* 12.2; Bac. *Intro.* 4.
2 Boet. *De inst. mus.* 1.8.195.2–5.
3 Bac. *Intro.* 10; Boet. *De inst. mus.* 1.8.195.6–8; Nicom. *Man.* 12.4.

interval meanwhile occurs, usually defined as follows. **5** A *diastema* is the difference or the interval of two phthongi which are dissimilar in height and lowness. **6** A system is said to be that which is harmoniously produced from more than two tones. **7** Therefore, the smallest of what is harmoniously sung is the phthongus. **8** But the smallest of these diastemata—distances or intervals—is the *diesis*. **9** The diesis is the minimum interval that our natural capacity can melodically lower and raise, **9a** according to Bacchius, about which more will be said in the third [chapter] of the fourth [book]. **10** The double of the diesis is the semitone. **11** The double of the semitone is the tone, as he asserts. **12** But the tone is also that by which the consonance of the diapente appears greater than that of the diatessaron—**12a** a fact that is treated more clearly in the fourth [book] of *The Theory of Music*. **13** Granted that an interval is the distance of a high to a low tone, as often as two dissimilar pitches are joined together, **14** either the interval will be larger, **14a** as, for example, the diapente [c4v], **14b** the diatessaron, and the diapason, **15** or it will be smaller, such as the ditone, the tone, the semitone, and the remaining conjunctive [intervals]. **16** Thus we are accustomed to ascribe a tone lying below and inferior to depth, but a superior one to height. **17** This Marcus Tullius [Cicero] in the sixth [book] of his *Republic* expressed in these words:

> **18** Nature, too, causes that the extreme intervals sound, from one part the low and from the other the high.

19 But if below a lower tone we join another, then the first lower tone will be distant in height from the other, **20** and it is made higher in comparison to the added one, while preserving its lowness. **21** The same thing occurs when above a higher tone another is added, **22** for the same reason, the first higher tone is made low in relation to the other, **23** not by lowering it, since it is preserved unmoved, **24** but by comparison to the other tone, which preserves its height. **25** Thus no particular, determinate, and stable position is established for height and lowness—**25a** which was asserted at the end of the

5 Bac. *Intro.* 6.
6 Bac. *Intro.* 5.
7–8 Bac. *Intro.* 7–8; see also Gaf. *De harm.* 1.2.
9 Bac. *Intro.* 8; cf. below 4.3.9–15c.
10 Boet. *De inst. mus.* 1.21.213.17.
11 Bac. *Intro.* 8.
12 Bac. *Intro.* 9; cf. below 4.4.
13 Boet. *De inst. mus.* 1.8.195.6–7; Nicom. *Man.* 12.3 ("An interval is the distance between two notes"). Gaf. refers to the same passage later in this chapter (cf. below 2.3.26).
17 Boet. *De inst. mus.* 1.27.219.12–13.
18 Boet. *De inst. mus.* 1.27.219.13–15; Cic. *Rep.* 6.18.
25a Cf. below 2.2.120–127, 142–146.

preceding chapter. **26** Hence it results that interval is the distance between high and low tones **27** and that consonance is the mixing of high and low tones. **28** Thus, we shall easily conclude that the division of the gamut into grave, acute, and superacute [as stated in] the *Introductorium,* is erroneous. **29** When a consonance occupies the range which [the theorists] consider only grave, it must be made up of graves and acutes **29a** for there to be a consonance.

30 Granted that the perception of the ears judges [consonance], the reason ponders it. **31** This occurs in this manner. **32** Whenever two strings, one tuned lower, struck at the same time, yield a somewhat mixed and pleasant tone, and the two tones concur in one, as if joined together, then a consonance is made. **33** Euphony is the same as consonance, for it is derived from "eu" which means good and from "phonia," which means fullness of tone, hence, good fullness of tone. **34** Yet Fabius [Quintilian] translates "euphonia" into Latin *vocalitas* [open sound, euphony]. **35** Augustine, however, translates *symphonia* as *concordia* [union] of sounds. **36** But *harmonia* with Hucbald is "the suitable uniting into one of diverse sounds." **37** On the other hand, Plato, in the second [book] of his *Laws,* as translated by Marsilio Ficino, states that harmonia is an order of the tones through a proper mixture of high and low,

26 Boet. *De inst. mus.* 1.8.195.6–7; Nicom. *Man.* 12.3; see also *Theorica* 2.3.13.

27 Boet. *De inst. mus.* 1.8.195.6–8 (cf. above 2.3.3).

28–29 Guido *Reg.* (cf. Gerbert, *Scriptores,* 2:25–29. On the use of "graves" and "acutae," see *Theorica* 5.4 ff.).

29a Boet. *De inst. mus.* 1.28.220.2.

30–32 Boet. *De inst. mus.* 1.28.220.2–7. For a more comprehensive discussion and definition of consonance, see André Barbera, "The Consonant Eleventh and the Expansion of the Musical Tetractys: A Study of Ancient Pythagoreanism," *Journal of Music Theory* 28 (1984):192ff.

33 March. *Lucid.* 3.2–3. In these generalizations Gaf. makes an implicit reference to Isid. *Etym.* 3.20.3f. (as cited in Fritz Reckow, "Diaphonia," *Handwörterbuch* (1971), pp. 2–3).

34 Quint. *Ist.* 8D.

35 March. *Lucid.* 5.5.2. The erroneous attribution of this definition to August. is found in the *Lucid.,* the most likely source for Gaf.'s adaptation in both his *Extractus* 7.1 and the *Theorica.* Further on the transmission of this information, see William Bowen, "St. Augustine in Medieval and Renaissance Musical Science," in: *Augustine on Music: An Interdisciplinary Collection of Essays,* ed. Richard R. LaCroix, vol. 6 of *Studies in the History and Interpretation of Music* (Lewiston, N.Y. and Queenston, Ontario: Edwin Mellen Press, 1988), pp. 38, 50.

36 Anon. *Mus. ench.* 9. (cf. Hans Schmid, ed., *Musica et Scolia enchiriadis,* p. 20). An attribution of the *Mus. ench.* to Huc. is contained in a number of manuscripts, with Manuscript Paris, Bibliothèque Nationale fons. lat. 7202 representing the oldest source for this attribution (cf. Hans Müller, *Hucbalds echte und unechte Schriften über Musik* (Leipzig: B. G. Teubner, 1884), pp. 3–33). I am indebted to Dr. Michael Bernhard (Bayerische Akademie der Wissenschaften, Munich) for this information. The *Mus. ench.,* although printed under Huc.'s name in Gerbert, *Scriptores,* 1:125 ff., is not the work of this author (cf. Lawrence Gushee, "Musica enchiriadis," *New Grove* 12:800–802). Incidentally, this statement does not occur in Huc.'s *De harm.* The erroneous attribution of this definition to Huc. is found in March.'s *Lucid.* 5.4.2, the most likely source for Gaf.'s adaptation.

37 Fic., trans. *Pl. Leg.* 2.795.

which we ourselves are used to call *concentus*. **38** The Philosopher pointed out that this relation of tones exists in the high and low. **39** Bacchius indicates that *melodia* is the [interplay of] slackening and tension made up of harmonious tones. **40** When two strings struck at the same time do not mix together in the ears as a pleasant and single tone but as if composed of two tones, this is called a dissonance. **41** Isidore declares **41a** that this dissonance mixed of two tones approaches the ears as a rough and unpleasant mixture. **42** They also call an aggregate of disharmonious tones *diaphonia*.

Chapter **Four**
The formation of consonance

1 Apollodorus [of Athens] asserts that sound from which consonance is generated occurs when air that is a medium between the speaker and the listener is beaten; **2** then, having been agitated, it flows into the ear in a circular motion just as, for example, the water in a cistern is set into circular motion when a stone has been thrown in. **3** [Also, Apollodorus asserts] that the sound itself occurs when the faculty of sensation is released around the originator in our mind. **4** Themistius, however, maintained that the air sounding initially does not itself travel and reach the hearing but sets in motion the air next to

39 Bac. *Intro.* 19, 45.
40 Boet. *De inst. mus.* 1.28.220.7–10. Boet. provides a more complete definition of consonance in *De inst. mus.* 1.8. Concerning Boet.'s theory of consonance, see also *De inst. mus.* 1.31.
41a Boet. *De inst. mus.* 1.8.195.8–10; Nicom. *Man.* 12.4; Isid. *Etym.* 3.20.3ff.
42 Though this statement lacks a reference to a source, it seems that Gaf. made an implicit reference to Isid. *Etym.* 3.20.3 ff. (as cited in Reckow, "Diaphonia," pp. 2–3).
1–3 The opinion ascribed here to "Apolodorus" comes from Diog. Laert. *Vitae* 7.158 (in *Stoicorum veterum fragmenta*, ed. Hans von Arnim, 4 vols. (Leipzig: 1903–24), 2: 234, no. 872) in his account of Stoic doctrine on hearing: where Gaf. has "rursus sonum quo consonantia generatur." Diogenes has ἀκοὺειν δέ. though von Arnim needlessly adds ἡμᾶς. No author is cited, but the construction is parallel to that in the immediately preceding account of vision (§157, *SVF* 2:233, no. 867) for which Diog. cites "Chrysippus in Book 2 of his *Physics* and Apollodorus" (sc. the late second century B.C. Stoic Apollodorus of Seleuceia-on-Tigris, modern Tell 'Umar). However, in the additional remark "sonum uero ipsum fieri, etc.," *sonum* must derive from a misprinting or misreading of *somnum*, for although the sentence (*SVF* 2:215, no. 766) follows on just as it does here. Diog. Laert. has changed the subject from sound to sleep: τὸν δέ ὕπνον γίνεσθαι κτλ. Gaf. was using the Latin translation of Diog. Laert. by Trav. (Sant'Ambrogio Camaldolese, 1386–1439), published in Rome, 1439. It is not clear by whose error "somnum" (or "Sõnum," as in Jenson's edition, Venice, 14 August 1475). This information was kindly communicated to the translator by Dr. Leofranc A. Holford-Strevens (Oxford, England) in letters of 13 July and 21 July 1992. For a more detailed explanation of this passage, see Kreyszig, *Franchino Gaffurio's "Theorica Musice")*, chapter 6, section 7.
4–6 Barb., trans. *Them. Paraphr. (De an.)* 2.29.85.

itself, **5** and the continuous [air] immediately after it excites other air in a fashion perceived in waves, **6** in which one [wave] pushes the other and the driven [wave] impels forward the following one. **7** For when a stone, thrown from a distance, sinks in a pond or calm water, a wave is at first concentrated in a very small circle and thereafter disperses the rounds of waves in bigger circles, **8** until the motion, exhausted from generating the waves, quiets down; **9** however, the wave is always diffused farther and wider.

10 But if a little obstacle is placed before the rising waves, this motion is instantly reversed **11** and is turned back with the same little waves almost towards the center from which it had set out. **12** So, too, in this manner, when the struck air makes a sound, the air strikes other air, and it excites a circular wave of the air, so to speak, **13** and in this manner is diffused and at the same time strikes the hearing of all situated around it. **14** Nevertheless, pitch is more indistinct to someone further away, because a weaker wave of the air that has been struck reaches him.

15 The Stoics affirm that sound itself is material, saying that sound is air that has been struck. **16** And the material body itself is that which either acts or is acted upon. **17** Lucretius, wishing to apply this principle, advances it by this reasoning **17a** that "nothing but a material body can touch or be touched." **18** Indeed, the Greeks say that a material body is that which exists in three dimensions. **19** Plato, however, believes that a sound is not material, as Aulus Gellius reports. **20** In fact, Plato says:

> the blow itself and the percussion, not the struck air, is the sound, **21** nor is merely a blow of air sound, **22** for a finger also strikes the air; **23** nevertheless, the finger does not thus produce a sound.

24 But Democritus and later Epicurus say that a sound is composed of individual particles.

25 Since the sounds are perceived by the ears, **25a** Aristoxenus for that reason decided to treat the consonances not according to ratio but leaves them to the judgment of the ear, **26** and he identifies these pitches by no proportions

16 Gell. *NA* 5.15.3. This statement is adopted from Pl. *Soph.* 247D.
17 Gell. *NA* 5.15.4.
17a Gell. *NA* 5.15.4; Lucr. *De rer.* 1.190.
18 Gell. *NA* 5.15.5–6.
19 Gell. *NA* 5.15.7; Pl. *Tim.* 67a.
20–24 Gell. *NA* 5.15.7–8.
24 Democr. *Guar.* (lost); cf. Epic. *Spic.*; cf. Hermann Usener, ed., *Epicurea* (Leipzig: B. G. Teubner, 1887).
25 Boet. *De inst. mus.* 5.13.363.5–7; Aristox. *Harm.* 2.32.
25a–26 Boet. *De inst. mus.* 5.13.363.5–9. *De inst. mus.* 5.13 is a condensed paraphrase of Ptol. *Harm.* 1.9–10, 19–24 (as cited below, 2.4.25–31).

of numbers. **27** Rather, he assigns the differences among these [pitches] to the medium [in which they occur], so that he orients the speculation not about the pitches themselves **28** but too shortsightedly around that which he identifies as the differences among them, **29** because he thinks that there is a difference between those pitches of which he sets up no image or measure. **30** He proposes that the diatessaron is a consonance of two tones and a semitone **31** and that the diapente consists of three tones and a semitone, and the diapason of six tones. **32** But this is wrong, **33** for we shall demonstrate that two smaller semitones [c5v] can in no way make a [whole] tone.

34 According to Plato, consonance comes about in the hearing in this way: **35** the higher of the two sounds, which is the speedier, precedes the low sound and enters the ear quickly, **36** and the innermost part of the ear, having been struck, bounces the impulses back, as it were, with repeated motion. **37** But now the sound emitted more slowly arrives, not as fast as its first impulse. **38** For this reason, the higher sound, now returning lower, presents itself as similar to the approaching low sound and is blended with it, making one consonance.

39 But this Platonic deliberation seems to disagree with Nicomachus, **40** for, as has been touched on earlier, **40a** not a resemblance but a difference [in sound] is responsible for this harmony; **41** for which reason, if a low sound is blended with [another] low [sound], it will not produce a consonance, **42** since consonance is a mixture of high and low sound, **43** by which means it happens that a difference separating individuals is coupled together in a blend. **44** Hence, Nicomachus, thinking that a consonance is made thus, says: **45** "It

27 Boet. *De inst. mus.* 5.13.363.9–11; Aristox. *Harm.* 1.15.
28–29 Boet. *De inst. mus.* 5.13.363.11–14.
30 Boet. *De inst. mus.* 5.13.363.14–16; cf. Gaf. *Pract.* 1.5.
31 Boet. *De inst. mus.* 5.13.363.16–17. This statement of Aristox. (*Harm.* 2.56–58) had already been proven wrong by Boet. (*De inst. mus.* 2.31; 3.3) in terms of Pythagorean mathematics; see also discussion in *De inst. mus.* 5.14, which is based on Ptol. *Harm.* 1.11.25–28. This fact accounts for the condensed presentation of the material in the present context of the *Theorica*. Boet. *De inst. mus.* 2.31 is paraphrased in *Theorica* 4.4. Furthermore, on the description of the diapente see Gaf. *Pract.* 1.6; on the diapason see *Pract.* 1.7.
32–33 See also below, 4.3, for a continuation of the discussion of this topic.
34 Boet. *De inst. mus.* 1.30.221.12–13; Pl. *Tim.* 80 A–B.
35–38 Boet. *De inst. mus.* 1.30.221.13–20.
39 Boet. *De inst. mus.* 1.31.221.22–23.
40 Gaf. here refers to the earlier passage in the *Theorica* 2.3.30–32, in recourse to Boet. *De inst. mus.* 1.28.220.2–7.
40a–43 Boet. *De inst. mus.* 1.31.221.23–26, 222.1–2.
44–46 Boet. *De inst. mus.* 1.31.222.2–6; an expansion of earlier statements in *De inst. mus.* 1.3.189.23–29, 190.1–7 (cf. above 2.1.36–47). Nicom.'s extant works do not discuss a theory of consonance; rather, Nicom. (in *Man.* 4.5) merely describes how one vibrating string is responsible for creating a multitude of sounds.

is not so much one impulse of air which emits a simple kind of sound, **46** but rather a string, once struck, drives the air very frequently and creates many sounds"—**47** doubtless because the velocity is that of the percussion, so that a sound in a certain way includes sound. **48** The distance between the sounds is not perceived, and it is as if a single sound reaches the ear. **49** If, therefore, those beatings of the low sounds are commensurable with beatings of the high, as in the proportions of the hammers, **50** there is no doubt that this commensurability of beatings mixed among themselves creates one consonance of sounds.

51 Our Boethius relates that since judging consonances belongs to the ears as well as to the reason, the reason ought to be the better judge of these. **52** For, indeed, in the same manner as the ear is affected by sounds or the eye by sight, so is the judgment of the mind affected by numbers or by continuous quantity. **53** If a number or line is presented, nothing is easier to judge with eye or mind than its duple. **54** Right after judging the duple follows the half, after the half the triple, after the triple the third part. **55** Since the form of the duple is simpler, however, Nicomachus considers the diapason to be the best consonance; **56** after this the diapente occupies the middle, then the diapason-plus-diapente, which is the triple, **57** and he judges the rest following the same procedure and form.

Chapter **Five**
In which proportions consonances consist

1 The consonances of music arise from joining height and lowness that are commensurate. **2** A measure is said to be common when the difference between two numbers measures both, as is evident in multiple proportions. **3** For that part which is the difference between two numbers measures what is

47–48 Boet. *De inst. mus.* 1.31.222.6–8.
49 Boet. *De inst. mus.* 1.31.222.8–10. This is also a reference to *Theorica* 1.8, see woodcut on fol. b6 recto (fig. 1.8.1).
50 Boet. *De inst. mus.* 1.31.222.10–12.
51 Boet. *De inst. mus.* 1.32.222.14–16; also 1.28.220.2–3 (cf. above 2.3.30).
52–54 Boet. *De inst. mus.* 1.32.222.16–22.
55 Boet. *De inst. mus.* 1.32.222.22–23; Nicom. *Man.* 5.1.
56 Boet. *De inst. mus.* 1.32.222.23–25.
57 Boet. *De inst. mus.* 1.32.222.25–26. Boet.'s attribution of the ranking of consonances to Nicom. is also mentioned in *De inst. mus.* 2.20. Yet such ranking cannot be substantiated from the extant works of Nicom.
1 Boet. *De inst. mus.* 1.29.220.12–14.
2 Boet. *De inst. mus.* 1.29.220.14–16. For a definition of the multiple genus, see Gaf. *Pract.* 4.3; also Tinct. *Prop.* 5.
3 Boet. *De inst. mus.* 1.29.220.16–18.

duple, as, for example, the difference between 2 and 4 is 2, which measures both. **4** Similarly, the same number [c6r], 3, measures 3 and 6. **5** And between 2 and 1 there exists the same unity measuring both; **6** between 9 and 6, the same number, 3, measures both. **7** Likewise, in superparticulars, if the sesquialter is a proportion like 6 and 4, the number 2 is what measures each, since it is the difference between them. **8** Given the sesquitertial proportion 8:6, the same number, 2, constitutes the difference and measures both.

9 But this does not happen in other genera of inequality, as in the superpartient, **10** for if we compare 5 to 3, 2, which is the difference, measures neither, **11** because 3 is too small and twice 3 is too large. **12** In the same manner, twice 2, compared to 5, does not reach 5; **13** thrice 2 exceeds 5. **14** For this reason this genus of inequality stands apart from nature and generation of consonances, **15** since in the first two genera [of inequality], which form consonances, there are many similarities, **16** but in this superpartient genus and the remaining two composite genera least of all. **17** This is proved in the following manner: **18** a duple is nothing but twice the simple [quantity], **19** and the triple is nothing else but three times the simple, **20** and the quadruple likewise is the same as four times the simple. **21** The sesquialter is three times one half, **22** the sesquitertial is four times the third, which is not found in other genera of inequality. **23** Now, since equal sounds [*aequivocae*] by comparison are close to unisons [*univoces*], it is necessary to join to equal numbers that inequality of numbers which is close to equal sounds. **24** Next to the equality of numbers is that proportion of inequality which is duple, **25** for it is the first species of the multiple genus, **26** and it occurs when the larger number exceeds the smaller by a number equal to the smaller number itself; **27** for example 2:1, for 2 exceeds 1 by 1, which is equal to the same number, 1. **28** Justly therefore, the duple proportion is assigned through equal sounds, as is easy to see, to the diapason. **29** Similarly, the bisduple or quadruple proportion is assigned to the bisdiapason. **30** The proportions dividing the duple proportion, however, must be adjusted to those consonances which divide the equal sounds of the

7 Boet. *De inst. mus.* 1.29.220.20–23. For a definition of the superparticular genus and the sesquialter proportion, see Gaf. *Pract.* 4.5; also Tinct. *Prop.* 6.
8 Boet. *De inst. mus.* 1.29.220.23–24. For a definition of the sesquitertial proportion, see Gaf. *Pract.* 4.5; also Tinct. *Prop.* 6.
9 Boet. *De inst. mus.* 1.29.220.24–26. On the five categories of proportions of greater inequality, see Gaf. *Pract.* 4.2; on the superpartient genus, see Gaf. *Pract.* 4.7; also Tinct. *Prop.* 7.
10–22 Boet. *De inst. mus.* 1.29.220.26–29, 221.1–10.
23 Boet. *De inst. mus.* 5.11.361.19–22; Ptol. *Harm.* 1.7.15–16. For a definition of *aequivocae* (=aequisonae), see above 2.2.132; for a definition of *univoces* (=*unisonae*), see above 2.2.131.
24–28 Boet. *De inst. mus.* 5.11.361.22–27; Ptol. *Harm.* 1.7.15–16.
29 Boet. *De inst. mus.* 5.11.361.27–28; Ptol. *Harm.* 1.7.15–16. For a definition of the quadruple proportion, see Gaf. *Pract.* 4.3.
30–31 Boet. *De inst. mus.* 5.11.361.28–31, 362.1–2; Ptol. *Harm.* 1.7.15–16.

diapason, **31** whereby it happens that the diapente is joined by comparison to the sesquialter and the diatessaron to the sesquitertial. **32** The consonances joined to equal sounds create other consonances, such as the diapason-plus-diapente in the triple proportion. **33** Accordingly, the diapason and the bis-diapason are equal sounding, since by their temperament and mixing they produce a single and simple sound.

34 Thus it is not unknown that the diapason in the duple proportion is the first proportion in the multiple genus; **35** the diapason-plus-diapente in the triple proportion is second; **36** the bisdiapason in the quadruple proportion is the third species of this multiple [genus]. **37** It is evident that the diapente is the first [species] of the superparticular genus, the sesquialter, **38** and that the diatessaron is contained in the second species, the sesquitertial. **39** But when the consonance of the diatessaron is added to a diapason, consisting of an equal sound, it does not produce a consonant sound; **40** nevertheless, Ptolemy demonstrated that it is a consonance not through the judgment of the ear but through reason in the following manner. **41** Because the consonance of the diapason creates such a conjunction of pitch, [c6v] it seems to be one and the same string [of a musical instrument], a fact with which the Pythagoreans also agree; **42** if whatever consonance is added to it, it is completed and unviolated. **43** Consequently, if a consonance of the diatessaron is joined to the equal-sounding diapason, it is as if the consonance of the diatessaron were joined to one tone. **44** On that account, Ptolemy contends that this happens because the diapason is nearly one small sound **45** and is a consonance of such a character that it is as if a single sound, just as a whole number added to 10 remains untouched and not violated. **46** This to the others does not happen; **47** if you add 2 and 3, they become 5, **48** and the species of number is changed; **49** if, however, you add these two to the number 10, you create 12, **50** and the 2, joined to the 10, has been preserved; **51** likewise the number 3, and the rest in the same manner. **52** Thus, the consonance of the diapason, whenever it takes on another consonance, preserves the consonance and does not change it, **53** and it does not yield a dissonance from the consonant sound. **54** Therefore, when the consonance of the diapente is added to an equal-sounding diapason, one creates the bringing-together of the consonant diapason-plus-diapente sounding in a triple proportion.

55 Thus the diatessaron, joined to an equal-sounding diapason, returns the other consonance from an embracing of the one to the other, **56** which,

32 Boet. *De inst. mus.* 5.11.361.2–3; Ptol. *Harm.* 1.7.16. For a definition of the triple proportion, see Gaf. *Pract.* 4.3.
33 Boet. *De inst. mus.* 5.12.362.10–12; Ptol. *Harm.* 1.7.16.
40–43 Boet. *De inst. mus.* 5.9.358.21–26, 359.1; Ptol. *Harm.* 1.6.13.
44–57 Boet. *De inst. mus.* 5.10.360.6–22; Ptol. *Harm.* 1.6.13.6D.

granted, does not result from a multiple or superparticular proportion, **57** in which that quantity, which is the difference between the two terms, is the common measure of both. **58** Rather it results from the superpartient genus, from the proportion 8:3 **58a** and the proportion 16:6, as is uncovered through the examination of the weight of the sixth hammer to the second. **59** Therefore the diapason-plus-diatessaron does not concord to the ear because the difference between the terms of the same proportion is not the common measure of both. **60** But we shall attribute this common measure to the other consonances with no difficulty; **61** that is, through a suitable addition of a tone to a diapente or of a sesquioctaval proportion to a sesquialter [proportion], another consonance grows to a rational consonant proportion. **62** Of course, to the extent that a tone is joined to a diapason-plus-diatessaron, it completes the consonance of the diapason-plus-diapente. **63** Thus the sesquioctaval proportion is added to the duple-superbipartient proportion, **64** that is, derived from the weight of the sixth hammer to that of the second; **65** this superbipartient proportion is raised to a triple proportion of the multiple genus. **66** On the other hand, if a diapente is added to a diapason-plus-diatessaron, it completes the consonance of the bisdiapason; **67** thus a sesquialter superimposed upon a duple superbipartient proportion of this sort will fulfill the quadruple proportion contained in the multiple genus. **68** Therefore we shall keep the diapason-plus-diatessaron, having observed that it nearly harmonizes, for it also complies with the simple sounding-together of the diatessaron.

Chapter **Six**
What number is, and its necessity in all things

1 Since the rationality of numbers inheres in all things created by primeval nature, this principle was an idea in the mind of the creator, **2** whence came the multitude of four elements, the alternations of the seasons, the motion of the stars, and the periodical return of the sky. **3** The state of all things [d1r] is governed by the ratios of numbers. **4** Certainly, it is necessary to discuss numbers themselves succinctly. **5** A number, according to Isidore [of Seville],

58 Boet. *De inst. mus.* 5.10.360.22–25. This is a reference to the myth of the Pythagorean hammers; see *Theorica* 1.8.
58a–60 Boet. *De inst. mus.* 1.10.196.18–28, 197.1–29, 198.1–8 paraphrased.
61 Boet. *De inst. mus.* 1.10.196.18–28, 197.1–29, 198.1–8 paraphrased. For a definition of the sesquioctaval proportion, see Gaf. *Pract.* 4.5; also Tinct. *Prop.* 6. For a definition of the rational proportion, see Gaf. *Pract.* 4.1.
62–68 Boet. *De inst. mus.* 1.10.196.18–28, 197.1–29, 198.1–8 paraphrased.
1–4 Boet. *De inst. arith.* 1.2.12.14–20.
5 Isid. *Etym.* 3.3 slightly expanded.

is a multitude assembled from units, a collection of units, **6** or, as Boethius asserts, a set of quantities consisting of units. **7** Certainly, were someone to remove number from human nature, he would not be thought wise; **8** nor will an animal which ignores 2 and 3, even and odd, and number altogether, ever be able to give an account concerning the things which it perceives through the sense alone and through the memory.

9 In all these things, the divine Plato in his *Philosopher* or in the *Epinomis* teaches that number itself is necessary, **10** which—although it can be shown through several reasons—is now correctly made clear through this principal reason. **11** For the other arts disappear completely if the concept of number is taken away, **12** and since we speak of our faculty of music, all of music itself requires number of motion and of sounds, **12a** from which rhythm is made, that is, order in time and motion. **13** Indeed, we can conveniently say that this rhythm is number. **14** In this way, we perceive number in time and motion. **15** Also, as [Marsilio Ficino], the translator of the *Epinomis* observed, Plato in many places not only praises arithmetic; **16** indeed, he wonders at it, **17** for he says that it spurs everyone to sharpen his ingenuity to the highest extent, **18** to strengthen his memory, **19** and to train his intellect to be most precise and quick for every speculation and action, **20** [and] that number has been handed down from God himself to mankind as a necessary instrument of the understanding. **21** When [number] is removed, the intellect appears senseless, **22** and the sciences and the arts disappear completely, **22a** as we have already said. **23** Plato proposes that everything is bad which lacks harmony, number, and beauty. **24** What is joined through agreeing numbers, however, is good.

25 "Number" [*numerus*] derives its meaning from *nummus* [coin], and it acquires the word by its frequent use. **26** "One" [*unus*] derives its name from the Greek. **27** The Greeks call *enna* what we call "one"; **28** likewise, we call "two" [*duo*] and "three" [*tres*] what they call *dya* and *tria*. **29** "Four" [*quattuor*] is said to be derived from *quadrata* ["square"]. **30** "Five" [*quinque*], however, has its name not according to nature but according to the whim of the man who first gave names to numbers. **31** "Six" [*sex*] and "seven" [*septem*] derive from the Greek; **32** for, in fact, in many words which have the rough breathing in Greek we put an "s" in place of the rough breathing. **33** Thus, *sex* stands for

6 Boet. *De inst. arith.* 1.3.13.11–12.
7–8 The comparison between the human being and the animal in terms of the recognition of numbers recalls an analogous statement of Pl. concerning the different perceptions of rhythm by the human being and by the animal (cf. Pl. *Leg.* 653d–654, 664e).
9–12 Fic., trans. *Pl. Opera* 699 (cf. Pl. *Epin.* 3.977d–e, 978a).
12a–22 Fic., trans. *Pl. Opera* 1006 (cf. Pl. *Epin.* 3.977d–e, 978a).
22a Cf. above 2.2.44 ff.
23–24 Fic., trans. *Pl. Opera* 845 (cf. *Argumentum* to Pl. *Epin.*).
25–45 Isid. *Etym.* 3.3.

hexa, septem for *hepta,* just like for the herb *herpillum* we say *serpillum.* **34** "Eight" is a borrowing from the Greek; **35** their *ogdo* is our *octo.* **36** *Novem* [nine] in our language is *nea* in theirs. **37** We call *decem* [ten] what they in fact call *deca*; **38** it has been called *decem* from the Greek etymology, where it binds and unites the interspersed smaller numbers; **39** *desmos,* according to them, is said to unite or bind. **40** Furthermore, "twenty" [*viginti*] is said to have been derived from "twice ten" [*decem bis*], "v" having been set in place of "b." **41** "Thirty" [*triginta*] is said to be generated from "three times ten" [*decenario ter*]; and thus without interruption up to "ninety" [*nonaginta*]. **42** "Hundred" [*centum*], however, is said to be a circle which is from "singing" [*cantus*]. **43** "Two hundred" [*ducentum*] is said to be "two times hundred" [*dya centum*]; in like manner, the remaining [numbers] up to a thousand [*mille*]; **44** *mille* certainly is from the word "multitude" [*multitudo*], from which the word "military" [*militia*] is derived, as if it were a multitude [*multitudo*]; **45** also *milia* ["thousand"] which the Greeks call *miriades,* with the letter changed; thus, by multiplication, this leads to infinity.

46 The science of these numbers is not at all to be condemned, inasmuch as it resides in many places of the sacred letters by an act of God. **47** Certainly, it has not been said vainly in praise of God: **48** "You have disposed all things well in weight, number, and measure." **49** For 6 [d1v] is the number which, as will be shown a little later, is perfect in its individual parts; **50** it proclaims the perfection of the universe through a certain expression of its own number. **51** And those forty days in which Moses and Elijah and our Savior himself fasted can by no means be understood without the disposition of numbers. **52** Indeed, other numbers have been drawn together in the sacred Scriptures, whose figures no one except those who know can solve. **53** Likewise, to us it has been given to a certain extent to insist upon the teaching of numbers, **54** when we learn from this discipline the hours, **55** when we discuss the courses of the months, **56** when we recognize the interval of the returning of the year; **57** to be sure, we are instructed through number, if we are not to be brought into disorder. **58** Were number removed, everything would perish, **58a** as was mentioned before in the third [chapter] of [Book] one: **59** "Take away com-

46–47 Isid. *Etym.* 3.4.
48 Isid. *Etym.* 3.4; Wisdom 11.21.
49–50 Isid. *Etym.* 3.4.
51 Isid. *Etym.* 3.4. Each biblical figure mentioned here refers to certain passages of the Bible: a) Moses: Exodus 24.18; b) Elijah: 1 Kings 19.4 and 8. Note that in the English Bible this reference is included in 3 Kings 19.4 and 8. The different location of the same passage results from the fact that the two Old Testament books of Samuel are commonly named by the Fathers of the Church the first and second books of Kings. Thus, 1 Kings in the Latin Bible corresponds to 3 Kings in the English Bible; c) the Savior: Matthew 4.2.
52–60 Isid. *Etym.* 3.4.
58a Cf. above 1.3.17 ff.

puting from the human race and blind ignorance will engulf all." **60** Nor can it be taken away from the other creatures, who ignore reason. **61** It is most certain that numbers are infinite; **62** for although, in any number and whatever multitude we believe we make an end by taking away, **63** we always add another number and again another and increase through multiplication to the infinite itself. **64** Art as well as practice and nature itself teaches this. **65** Indeed, each number is so terminated by its properties that one cannot be equal to another, **66** and consequently, they are uneven among themselves and different, **67** and individual ones are finite. But all together they are infinite; **68** they are made in order that separate quantities may be distinguished.

Chapter **Seven**
The diverse classification of numbers

1 The classification of numbers is handed down by the arithmeticians in the following way. **2** They say that one number is even, another is odd. **3** Of even numbers, one is evenly even, another evenly odd, and another oddly even. **4** Of the odd numbers, one is prime and simple, another secondary and complex, **4a** and another is medial. **5** An even number is one which can be divided into two equal parts, **5a** for example, 2, 4, 8, and 10; **5b** or, as Boethius includes in his *On Arithmetic* as an example: **6** an even number is one which differs from an odd number by one, either by augmentation or diminution, **7** for if we add or subtract one from an odd number, it is made even. **8** An odd number is one which cannot be divided into two equal parts, since the middle is either lacking or superfluous, **8a** as for example 3, 5, 7, 9, **8b** and the like. **9** An odd number is one which differs by one from an even by augmentation or diminution, **10** for if we subtract one from or add it to an even number, an odd one is immediately created. **11** An evenly even number is that which divides

2–61 Isid. *Etym.* 3.5. The order of presentation in *Theorica* 2.7 seems to suggest strongly that Gaf. followed Isid. *Etym.* as his primary source of information rather than Boet. *De inst. arith.*, which he cites in the body of his text. Here Gaf., either deliberately or intentionally, failed to acknowledge his principal source, perhaps to conceal his affinity to Isid.'s writing. Like Boet., Isid. is also indebted to Nicom. *Intro. to Arith.* 1.8–16.
2–3 Boet. *De inst. arith.* 1.8.17.2–7.
4 Boet. *De inst. arith.* 1.13.30.9–11. Gaf. substitutes "simplex" for Boet.'s "incompositus."
4a Boet. *De inst. arith.* 1.13.30.11–13 paraphrased.
5 Boet. *De inst. arith.* 1.3.13.13–14.
6 Boet. *De inst. arith.* 1.6.15.23–24.
8 Boet. *De inst. arith.* 1.3.13.14–16 paraphrased.
9–10 Boet. *De inst. arith.* 1.6.15.24–28.
11–16 Boet. *De inst. arith.* 1.9.17.9–17; Nicom. *Intro. to Arith.* 1.8.

evenly by another even number, [d2r] through which it always attains an indivisible one, **12** as for example, 64, whose middle is 32, **13** the middle of which is 16; **14** furthermore, the middle of this [16] is 8, the middle of which is 4; **15** and the middle of 4 is 2, **16** whose indivisible unity preserves the middle. **17** An evenly odd number is one which accepts a mathematical division into equal parts, **18** whose individual parts in two equal parts stay indivisible, **18a** as for example, 6, 10, 14, **18b** 30, and 50. **19** An oddly even number is that of which parts can also be divided, **20** but [these parts] do not arrive at a unit, as for example, 24, **21** which, divided in the middle, produces 12, **22** which again divided evenly yields the number 6, **23** the middle of which indeed is 3, **24** which does not yield halves; **25** hence it does not reach the indivisible unit one. **26** Oddly odd is a number that is measured by an odd number, as for example, 25 and 49, **27** which are uneven because they are divided by odd numbers, like five times 5 and seven times 7, respectively. **28** Indeed, of the odd numbers, some are simple, others are composite, and others, **28a** as said previously, **28b** are medial. **29** Simple numbers are those which have no aliquot parts other than unity, **30** as for example, 3, of which 1 is the third part, and 5, which has only a fifth part, that is, the unity itself, **31** and 7 only a seventh, **32** since for these numbers unity itself is the only aliquot part **32a** and the common measure of all numbers.

33 Composite [numbers] are those which are not measured by a single unit but by some other aliquot parts as well, **34** as, for example, 9, which is reduced by aliquot parts of 1 and 3; **35** the number 15, having aliquot parts of 1, 3, and 5. **36** Likewise, 21 has as aliquot parts unity itself, 3, and 7.

37 Medial numbers are those which in a certain way seem to be simple and

17–18a Boet. *De inst. arith.* 1.10.21.24–26; 22.1–4; Nicom. *Intro. to Arith.* 1.9.
19–25 Boet. *De inst. arith.* 1.11.25.5–16; Nicom. *Intro. to Arith.* 1.10.
26–27 Boet. does not comment on the oddly odd number ("numerus impariter impar") but instead concentrates on a more detailed exposition of the oddly even numbers (cf. Boet. *De inst. arith.* 1.12.28.18–26, 29.1–28, 30.1–3). Nicom., too, does not mention this classification of numbers in this terminology.
28, 28b Boet. *De inst. arith.* 1.13.30.5–14 summarized.
28a Cf. above 2.7.4a.
29–32 Boet. *De inst. arith.* 1.14.30.16–19, 31.9–15. Note the slight difference in terminology. Gaf., though he adopts the reasoning of Boet., does not employ Boet.'s terminology of "primus et incompositus" but instead he uses the term "simplex" (and later also the term "incompositus") to describe the same facts. Boet. adopted the terminology from Nicom. *Intro. to Arith.* 1.11.
32a Boet. *De inst. arith.* 1.14.31.9–15.
33–36 Boet. *De inst. arith.* 1.15.31.24–30, 32.1–11 condensed; Nicom. *Intro. to Arith.* 1.13.
37–40 Boet. *De inst. arith.* 1.16.32.24–26, 33.1–15 paraphrased; Nicom. *Intro. to Arith.* 1.13. Gaf.'s terminology "mediocris numerus" is not found in Boet., whose description of this class of numbers does not resort to a specific term but merely points out the properties of this class of numbers.

incomposite and yet are composite in another way. **38** For example, when 9 is compared to 25, it is prime and incomposite, because it does not have a common number, except a monad, namely unity. **39** If, however, the same 9 is compared to 15, the number [9] is secondary and composite, because they have a common number other than the unit, **40** namely 3, which taken three times, makes 9, and 3 taken five times makes 15. **41** And again, they say that of the even numbers, some are excessive, others deficient, and others perfect.

42 Excessive numbers are those for which the sum of its aliquot parts exceeds the total number itself, as for example, 12. **43** A part is said to be aliquot if it yields the total precisely, when multiplied several times, **44** and indeed, some part can be said to be the middle, or third, or fourth, and so on. **45** The aliquot parts of the number 12 then are **46** 1, which is the twelfth part, **47** and 2, [d2v] which is its sixth part, **48** and 3, which is its fourth part, **49** and 4, which is its third part, **50** and 6, its half, **51** which, when added together, are shown to exceed 12 itself, of which they are individual parts, **52** since 1, 2, 3, 4, and 6 total the number 16.

53 Deficient numbers are those whose aliquot parts do not complete it fully, **54** for example, 8, because it has 1 as its eighth part, 2 is its fourth part, and 4 is its half, **55** which, when added together, do not arrive at 8, **56** since 3 is not said to be an aliquot part of 8, because it cannot be called a part of it.

57 Perfect numbers are those whose parts added together, yield the total precisely, **58** as for example, 6, whose sixth part is of course 1, the third part obviously 2, and half obviously 3, **59** which when joined, complete the number 6 precisely. **60** Within the first ten numbers, the number 6 is perfect; **61** within the first hundred numbers, the number 28 is perfect; **62** within the first thousand, 496. **63** Certainly, Saint Ambrose venerated the perfection of the number 6 in his *Hexameron* with incredible respect, **64** and Augustine in the eleventh book of his *City of God* is shown to have glorified it by a similar opinion in these words:

> **65** However, on account of the perfection of the number 6, this perfect repetition is narrated six times in 6 days; **66** not because the delay of time

41 Boet. *De inst. arith.* 1.19.39.20–22. Noteworthy here, too, is the slight deviation in terminology between the authors. Boet.'s "secundum perfectos, imperfectos," and "ultra quam perfectos" (as given in the heading of Boet.'s Chapter 18) corresponds to Gaf.'s "perfectos, diminutivos," and "superfluos." In this instance, Gaf. improves Boet.'s text by eliminating the inconsistency in terminology.

42–52 Boet. *De inst. arith.* 1.19.39.27–28, 40.1–9 paraphrased; Nicom. *Intro. to Arith.* 1.14.

53–55 Boet. *De inst. arith.* 1.19.40.11–23 paraphrased; Nicom. *Intro. to Arith.* 1.15.

56 Boet. *De inst. arith.* 1.19.40.11–23 paraphrased; Nicom. *Intro. to Arith.* 1.15.

57–59 Boet. *De inst. arith.* 1.19.41.6–20 paraphrased; Nicom. *Intro. to Arith.* 1.16.

60–62 Boet. *De inst. arith.* 1.20.42.3–5; Nicom. *Intro. to Arith.* 1.16.

63 Ambr. *Hex.* 6.1; 6.10.

65–69 August. *De civ. D.* 11.30.

appeared necessary to God, **67** as if he could not have created everything at once, **68** which would in turn by regular movements accomplish [the successive units of] time, **69** but since the perfection of the work has been sealed by means of the number 6.

70 Of these [numbers], 6 is certainly the first which is completed by its parts, as has been demonstrated, **71** whereby likewise Moses himself declares that the highest creator [God] perfected all his works.

Chapter **Eight**
The principles of numbers and their progressions

1 We are accustomed to call that which cannot be anything but the beginning of something a principle. **2** Likewise that which we call the end is that which cannot be anything but the end of something. **3** But one can arrive at an end from a beginning only through some middle. **4** Whence it arises that something is a whole in so far as it consists of a beginning, a middle, and an end. **5** Now the beginning, middle, and end are contained in the number three. **6** On that account a certain perfection is observed in this [number], because it contains the beginning, the middle, and also the end.

7 Unity alone is the beginning [of the series of all positive integers], because it lacks middle and end; **8** certainly unity is no number but the first [principle] of all numbers that are naturally disposed. **9** It is rightly—albeit in a long process—acknowledged as the mother of all plurality [d2r] **9a** since, in fact, it is the one which is the creator and mother of numbers. **10** Further, unity is a common measure of [numbers], **10a** what our Boethius in his *On Arithmetic* and Augustine in the first [book] of his *On Music* seem to indicate more clearly.

11 [The number] 2 naturally ensues as the first number different from unity, for the reason that it is disjoined from unity; **12** and therefore, it is the beginning of a certain otherness, because it is different only by a unity from the first and unchanging substance. **13** For the number 2 arises from two unities, which are two beginnings. **14** On the other hand, the first unity is a beginning from nothing, **15** the second unity is a beginning from the first: 1

71 Gaf. here refers to two passages from the Bible, Genesis 1.31 and Exodus 20.11.
1–6 August. *De mus.* 1.12.20.
7 Boet. *De inst. arith.* 1.20.45.1–2; 2.6.91.6–8; *De inst. mus.* 2.3.10–12.
8–9 Boet. *De inst. arith.* 1.7.16.23–26.
9a–10 Boet. *De inst. arith.* 1.18.37.17–18.
10a Boet. *De inst. arith.* (cf. 9a–10); August. *De mus.* 1.12.21 paraphrased.
11–12 Boet. *De inst. arith.* 2.27.117.5–10.
13–20 August. *De mus.* 1.12.21.

and 1 are 2. **16** And both of these are beginnings, **17** with the result that all numbers are indeed derived from the one, because they are made through multiplication and also through a kind of addition. **18** Yet the origin of multiplication and addition is assigned to a dual number, **19** and this becomes the first beginning **20** and, indeed, the other beginning through which all numbers are invented.

21 These two beginnings, that is, 1 and 2, having been joined with themselves, form the whole and perfect number. **22** For in counting, after 1 and 2 we place 3. **23** Consequently, this number, which is made from 1 and 2, is established after both in such order that no other can be interposed, **23a** which happens not at all in other numbers. **24** For if we unite two naturally neighboring numbers, the number arising from both, coupled in this manner, does not follow them in their natural order, since another number is interposed, **25** so that when we add 3 and 2, 5 is thereupon generated, which does not follow them directly, **26** but the interposed 4 occupies this natural continuation. **27** So 3 and 4 constitute 7, which does not follow them directly, for between 4 and 7 naturally lie 5 and 6. **28** The farther we try to progress in this manner to the more remote numbers, the more numbers are interposed. **29** So the great union is in the first three numbers, which we call these, that is, 1, 2, and 3, since nothing can be interposed between them. **30** From 1 plus 2 is made 3, which succeeds directly, for indeed, 1 plus 2 are 3.

31 If we proceed from 1 to 4, then a very true progression will be made because of [the presence of both] an odd and even number, **32** since the first whole odd number is 3, and the first whole even number is 4. **33** 1 and 2, then, are beginnings, and so to speak, the seeds of the numbers, from which this 3 is made, **34** so that now there are three numbers which, when joined to each other as if by a logical proposition, 4 appears and is produced. **35** They are rightly joined with these numbers so that a moderate progression is made to that number.

36 The number 5 is coupled in a natural mixture, as Martianus [Capella] asserts. **37** For it is established that 5 is comprised of a number from each sex, for 3 is believed to be male, whereas 2 is regarded as female. **38** Granted that [d3v] Martianus is owed respect as an interpreter of the individual and simple numbers, **39** nevertheless it is established that these first four numbers have a more honorable series and connection than the other numbers, **40** because

21–23 August. *De mus.* 1.12.22.
23a–24 August. *De mus.* 1.12.22 paraphrased.
25–30 August. *De mus.* 1.12.22.
31–35 August. *De mus.* 1.12.25.
36–37 Cap. *De nup.* 7.735.8–11.
38 Cap. *De nup.* 7.731–759.

these four numbers, when they are assembled in one, are known to complete the number 10, because 1, 2, 3, and 4 added together make 10. **41** Also in this number 10, which has the greatest value with respect to all the rest, are constituted certain points or infinite numbers, as when progressing from 1 to 10, those who count must descend again to 1. **42** In like manner we comprehend all these consonant proportions of these four numbers in series, whose process it is our present work to derive, **43** for, from these numbers, the diligent investigator will find out the duple, triple, and quadruple, sesquialter, and sesquitertial proportions.

BOOK **THREE**

Chapter **One**
Discrete and continuous quantity

1 Mathematicians assert that quantity is twofold, that is, continuous and discrete. **2** They have said that a quantity which is continuous is connected in all its parts and is not distributed in any limits, **3** as are, for example, a line, a stone, a circle, and all bodies, which are assigned their own particular magnitude. **4** For magnitude is the name given to that continuous quantity which begins from a bounded quantity **4a** and proportionally retains a section unlimited [as to proportion], **5** thereby losing the quantity of its measure as it increases, and diminishing as it multiplies. **6** For the line that is continuous is made ever smaller by infinite partitioning. **7** Suppose it is ten cubits in length, and this is divided at the exact half, **8** then that half is cut into another half, **9** and likewise, that half is divided into another half. **10** When this process is repeated ad infinitum, the quantity will always be found to be smaller, a property and nature the superparticular proportion seems to preserve, **11** since the proportion reduces the smaller number indefinitely in the larger one, as one observes with either its half or the third, or the fourth. **12** Indeed, the part named after the larger number is found to be smaller than the part of the smaller number, **13** since the third part comes from 3, as also the fourth part from 4. **14** And since 4 is larger than 3, it is necessary that the fourth be smaller than the third. **15** Certainly, discrete or disjunct quantity is that which, delimited by its parts [d4r], is reduced into one unity as if by accumulation, **15a** as Boethius has proposed in the introduction of his *On Arithmetic*.

16 What is called multiple exists when many parts, separated from each other at their extremes, are brought together into one. **17** Also, beginning from unity, than which nothing is smaller, one proceeds to infinity multiplying the increments, **18** and no number ever reaches the limit of growing. **19** Therefore, since it is the nature of numbers to progress to infinity and to increase, the multiple seems to serve the nature of number best, since it has no limit of growth. **20** Surely, the superpartient [proportion], which deviates somehow from simplicity

2–3 Boet. *De inst. arith.* 1.1.8.15–18.
4 Boet. *De inst. mus.* 2.3.228.6–7; Boet. *De inst. arith.* 1.1.8.18–19.
6 Boet. *De inst. mus.* 1.6.193.16–17.
15 Boet. *De inst. arith.* 1.1.8.19–20.
16 Boet. *De inst. arith.* 1.1.8.21–23 altered.

and retains two, three, or four parts and abounds in a certain plurality of these parts, neither preserves the whole nor takes away from the individual parts, **21** for which reason the Pythagoreans agree that this genus is not at all fitting for musical harmony. **22** The multiple is preserved in wholeness of numbers, for a duple holds the whole minor term twice, a triple three times, a quadruple four times, and so for the remaining numbers. **23** Superparticularity does not preserve any complete whole, however, but, in fact, exceeds it either by a half, by a third, or by a fourth, and so on. **24** Nevertheless, superparticularity produces division in single and simple parts. **25** Thus the properties of these quantities are opposite, for the one property, when multiplied, decreases in quantity, while the other superparticularity is shown to acquire an increase in quantity. **26** But of the multiple, which is called discrete quantity, some appear to stand by themselves, such as 1, 2, 3, 4, and so on. **27** And some are said to be related to another, such as the duple, the triple, the quadruple, and so on. **28** In fact, every number which requires nothing else to exist by itself is said to be simple and to exist by itself, such as 2, 3, 4, 5, and 6. **29** But that number which does not exist of itself, requiring another number so that it may exist, is said to be related to another number, such as the duple, the triple, the sesquialter, the sesquitertial, and so on, which are allotted denomination by their relation to one another. **30** For indeed, arithmetic integrity examines this multiple which exists in itself, **30a** that is, simple discrete numbers, **30b** in [Boethius's] *On Arithmetic*. **31** But musicians are familiar with that multiple which is related to a number as degrees of melody.

32 Of the quantity of magnitude **32a** that is said to be continuous, some numbers are stable and void of motions. **33** Geometry provides knowledge of this continuous quantity. **34** Others revolve continually in rapid rotation, at no time coming to rest. **35** The discipline of astronomy has laid claim to this knowledge. **36** Therefore, we are forced to address ourselves to discrete quantity, which is related to something with particular relevance to music in the present investigation.

Chapter **Two**
The definition and distinction of proportions

1 Of proportions which are formed by the mutual relation of numbers, one is called "proportion" in a common sense, another in a special sense. **2** A

30, 30b, 31, 35 Boet. *De inst. arith.* 1.1.9.2–6.
1 Ugol. *Decl.* 4.4.2. The absence of Ugol.'s name in both the margin and the text does not necessarily rule him out as the principal source for this chapter; in fact, the omission here is clearly misleading. The train of thought and the language of Chapter 2 leave no doubt as to Gaf.'s source.
2 Ugol. *Decl.* 4.4.2. In this passage Ugol. shows his indebtedness to Arist. *Top.* 1.15. Presumably, the reference here is to Boet.'s lost commentary on Arist. *Top.*—one of three

proportion is usually designated as the configuration of two things compared to each other in any univocality and is especially to be designated of univocal things, **3** because among equivocal things no comparison can really be found, according to Aristotle in the first [book] of [his] *Topics*. **4** Equivocal things are said to be those who have a name in common, **5** but the nature of the substance [i.e., the concrete individual thing] is diverse. **6** However, univocal things are said to be those whose names are common, **7** and also the nature of whose substance is the same. **8** Therefore, a voice and a stylus cannot be compared in sharpness **9** because, granted that the voice may be "sharp" and the stylus, too, **9a** nevertheless "sharpness" is not used univocally of both; **10** rather it has an equivocal meaning. **11** Certainly, both do not have the same nature but different natures, **12** whence it happens that between any equal things and unequal things whatsoever, **13** and between like and unlike, this proportion, as it is commonly called, can be ascertained, because it is called a "common" proportion. **14** On the other hand, a proportion meant in a special sense is the mutual relation of two quantities of the same related genus. **15** Hence we do not mutually compare here the continuous and discrete quantities—although the quantities are contained below the same genus of quantity—**16** since they are not neighboring but distant, **16a** as has been shown in the definition of these [genera of proportions]. **17** Indeed, from this description, this can easily be gathered, namely, **18** that proportion properly called, in a special sense occurs in those quantities which accept equality.

19 Of these proportions, some are rational, others irrational. **20** A rational proportion is a ratio of two commensurable quantities. **21** On the other hand, an irrational proportion is a ratio of two incommensurable quantities. **22** Commensurable quantities are said to be those of which there is one common measure, that is, forming any one of these quantities exactly, **22a** as for example, a two-foot length and a three-foot length, **23** for one foot taken several times gives each of these lengths exactly; **24** this is so because there is a measure common to both (unity). **25** The same applies to numbers, such as, 4 and 6, for 2 is their common measure [greatest common divisor]. **26** This also

---

works by Boet. on the *Top.* (cf. Eleanore Stump, "Boethius's Works on the Topics," *Vivarium* 12/2 (1974):81).

3 Ugol. *Decl.* 4.4.3.

4–5 Ugol. *Decl.* 4.4.4. Ugol.'s source here presumably is Boet., trans. Arist. *Cat.* 1.3–5.

6–7 Ugol. *Decl.* 4.4.4. Ugol.'s source here presumably is Boet., trans. Arist. *Cat.* 1.9–10.

8–16 Ugol. *Decl.* 4.4.4–6.

16 Cf. above 3.1.

17–20 Ugol. *Decl.* 4.4.7–8.

21 Ugol. *Decl.* 4.4.9.

22 Ugol. *Decl.* 4.4.10. The "quantitates commensurabiles" are also discussed in Boet. *De inst. arith.* 1.18.

22a–24 Ugol. *Decl.* 4.4.10.

happens in musical notation, **27** for if we take the long and the breve, the semibreve, which is the common measure of these, measures both. **28** We call incommensurable such quantities as have no common measure that yields any of these quantities exactly, **28a** for example, the diameter [i.e., the diagonal] of the square and its side. **29** For given a quantity which taken several times yields the diameter exactly, this same quantity or one equal to it taken several times never will yield the side exactly but more or less. **30** The reverse is also true, when reducing the side by this quantity. **31** In numbers, the superpartient proportions work the same way, as for instance, 5:3, 8:5, and the others in which there is no common measure.

32 Further, some proportions are said to be of equality, others of inequality. **33** The proportion of equality consists of a ratio of two equal quantities [d5r], **33a** such as 2:2, 3:3, and 1:1. **34** Equality exists when something compared with something else neither exceeds it by being a larger amount, nor is it exceeded by it by being a smaller amount. **35** In fact, this equality is naturally indivisible. **36** No one will say of equality that one thing is such and such and the other is of another type, since every equality preserves one measure in its own regularity. **37** For this reason that which is compared and that to which it is compared will always be called by the same designation, **38** so that if a 2 has been related to [another] 2, both are called 2.

39 Proportion of inequality consists of a ratio of two unequal quantities, such as 2:1, and of 1:2. **40** Inequality is that which, when a number is compared to another, either surmounts it by being a larger amount or is exceeded by it because of it being a smaller amount. **41** To be sure, an unequal thing is divided into a larger and a smaller, which are joined to each other in an opposite denomination [i.e., nomenclature]. **42** Certainly, the larger amount is larger than the smaller amount, and the smaller is smaller than the larger. **43** Nevertheless, both are not designated by the same words, as was said of equality, but by diverse and distinct words.

44 The science of music considers only this proportion of inequality, **44a**

27–30 Ugol. *Decl.* 4.4.11 paraphrased.
33–33a Ugol. *Decl.* 4.4.15.
34 Boet. *De inst. arith.* 1.21.45.13–15; Ugol. *Decl.* 4.3.2. Here the process of transmission described earlier in this chapter is reversed. The syntax and choice of vocabulary points to Boet. as Gaf.'s principal source. Further, this passage shows Ugol.'s indebtedness to Boet.; the affinity is already evident in Ugol.'s adaptation of Boet.'s chapter heading. Nicom. *Intro. to Arith.* 1.23.4; 2.1.2.
35–38 Boet. *De inst. arith.* 1.21.45.18–23; Ugol. *Decl.* 4.3.2 (cf. also the remark to 34); Nicom. *Intro. to Arith.* 2.1.2.
39–39a Ugol. *Decl.* 4.4.16.
40 Ugol. *Decl.* 4.3.3.
41–43 Boet. *De inst. arith.* 1.21.45.26–30; 46.1; Ugol. *Decl.* 4.3.3–4.
44–46 Ugol. *Decl.* 4.3.5.

since consonances are derived from unequal and dissimilar sounds, **45** and it is necessary that these sounds be produced by unequal terms and proportions. **46** In light of this kind of relation, we call the terms of comparison "numbers." **47** Therefore I consider it worthy of note that the origins of inequality proceed from equality in the same manner **47a** as arithmeticians say that all numbers have grown and originated from unity, **47b** which is not a number but the beginning of number. **48** This our Boethius has set forth more distinctly in his *On Arithmetic* and also in his *On Music*. **49** We, too, demonstrate this in the final [book].

50 Of the unequal proportions, some are called proportions of greater inequality, others of smaller inequality. **51** A proportion of greater inequality is a ratio of a larger quantity to a smaller, such as 2:1, 4:2, and 5:3, and similar [proportions]. **52** A proportion of smaller inequality is a ratio of a smaller quantity to a larger, such as 1:2, 2:4, and 3:5. **53** In these proportions of greater and smaller inequality there occurs a certain excess or difference which is that quantity by which either the larger number surmounts the smaller, or by which the smaller number is surpassed by the larger; **54** as when we compared 6:4, when 2 is the excess or difference; similarly if we were to adjoin 4 proportionally to 6.

Chapter **Three**
The five genera of proportions of greater and smaller inequality

1 A proportion of greater inequality occurs whenever a larger related to a smaller number comprises in itself this smaller number either several times exactly, **2** or only once and an aliquot part of the whole, **3** or only once and, besides, several aliquot parts, **4** or several times plus an aliquot part, **5** or several times and several aliquot parts. **6** Thus there are five genera of proportions of this kind. **7** The first is called multiple, the second, superparticular,

47, 47b Boet. *De inst. arith.* 1.32.66.18–22; Boet. *De inst. mus.* 3.5.276.15–18 glossed; Nicom. *Intro. to Arith.* 2.7.3; cf. also above 2.8.8.
47a Boet. *De inst. arith.* 1.32.66.16–22; Boet. *De inst. mus.* 3.5.276.15–18 glossed.
48 Boet. *De inst. arith.* 1.32.66.18–22; Boet. *De inst. mus.* 3.5.276.15–18 glossed.
49 *Theorica* 5.4.
51–53 Ugol. *Decl.* 4.4.17–18.
6–9 Ugol. *Decl.* 4.5.2. Although Gaf. does not acknowledge Ugol. as the principal source for his information, the wording here as well as in the subsequent passages clearly suggests a close affinity. In fact, Gaf., for the most part, quotes his authority almost verbatim. Most passages, however, can also be found in Boet., so that a clear pattern of transmission emerges from Boet. via Ugol. to Gaf. Boet. *De inst. arith.* 1.22.46.6–9; Nicom. *Intro. to Arith.* 1.18.19–23.

the third, superpartient, **8** and these three are the simple genera. **9** The fourth is called multiple-superparticular, and the fifth, multiple-superpartient. **10** To these five genera, five others of smaller inequality are recognized to be opposed, **11** and they are named merely by [the addition of] the preposition "sub." **12** They are the submultiple, the subsuperparticular, the subsuperpartient, the submultiple-superparticular, and the submultiple-superpartient.

13 Now the multiple genus is that which is more outstanding and more worthy and also far more ancient than the others. **14** This happens when a larger number compared with a smaller contains in itself this entire smaller number precisely several times, **14a** as twice, three, or four times, so that nothing is lacking and nothing exceeds. **15** The species of this genus are infinite. **16** The first species is the duple proportion, which is made when the larger term compared with the smaller contains in itself this smaller term twice, **16a** for example, 2:1, 4:2, 6:3, 8:4, and similar [proportions]. **17** For 2 comprises in itself 1 twice, and 4 similarly possesses 2 twice. **18** Also 6 possesses 3 twice, and 8 possesses 4 twice. **19** Nothing is lacking, and nothing is superfluous. **20** The triple proportion is the second species of the multiple genus; **21** it occurs whenever a larger number compared to a smaller one includes this smaller number precisely three times, for example, 3:1, 6:2, 9:3, **21a** and 12:4, **22** for 3 contains 1 three times, similarly 6 contains 2 three times, and 9 contains 3 three times exactly. **23** Also 12 comprises 4 **23a** precisely three times. **24** The quadruple proportion, which is the third species of this multiple genus, occurs when a larger number compared with a smaller one encloses in itself this entire smaller number precisely four times, for example, 4:1, 8:2, 12:3, and so on. **25** The remaining species of this genus proceed from the

10–11 Ugol. *Decl.* 4.5.3–4; Boet. *De inst. arith.* 1.22.46.9–17.
12 Ugol. *Decl.* 4.5.4; Nicom. *Intro. to Arith.* 1.18.2–3.
13 Ugol. *Decl.* 4.6.2; Boet. *De inst. arith.* 1.22.46.19–21; see also below 3.4.25.
14 Ugol. *Decl.* 4.6.2; Boet. *De inst. mus.* 1.4.191.11–12; Nicom. *Intro. to Arith.* 1.18.1.
14a Boet. *De inst. mus.* 1.4.191.12–13; Nicom. *Intro. to Arith.* 1.18.1.
15 The discussion of the species of the multiple genus is continued in 3.4.3–7. The property of infinity of this genus is shown in the table provided in *Theorica* 3.4.7a. The multiple genus is further discussed in *Theorica* 3.5.2–12.
16 Ugol. *Decl.* 4.7.2; Nicom. *Intro. to Arith.* 1.18.4.
19 Boet. *De inst. mus.* 1.4.191.13.
20–21 Ugol. *Decl.* 4.10.2; Nicom. *Intro. to Arith.* 1.18.5.
22, 23a Ugol. *Decl.* 4.10.2.
24 Ugol. *Decl.* 4.13.2; Nicom. *Intro. to Arith.* 1.18.6.
25 With regard to the compound ratios of the respective species, the numerator is also raised to infinity, e.g., 4:1, 8:2, 12:3, etc. Given the subsequent account of the same material in *Theorica* 3.4, this concept of the infinity of progressions is illustrated further through an arrangement of the prime form of each species in tabular form

(i.e., 2 3 4 5 etc., see below 3.4.7a).
1 1 1 1

multiple comprehensiveness of the smaller number, raising the denominator up to infinity.

26 A genus is said to be superparticular when a larger term related to a smaller one contains in itself this smaller term only once and in addition a single aliquot part of it. **27** But if that aliquot part by which this smaller number is surpassed by the larger number is a half, the proportion is called [d6r] sesquialter, **27a** for example, 3:2, 6:4, **27b** 9:6, and so on. **28** If, however, this part is a third of the smaller number itself, the proportion is called sesquitertial, for example, 4:3, 8:6, 12:9, and so on. **29** There are infinitely many species of this genus. **30** The first species is the sesquialter, the second the sesquitertial, the third the sesquiquartal, the fourth the sesquiquintal, and so on. **31** Every superparticular proportion, then, derives its designation from "sesqui," and from the aliquot part of the smaller number comprehended by the larger **32** so that if I form a sesquialter [proportion] such as 3:2—the one [number] 2 is equal to the other 2, but to one of them is added 1, which is the half of both 2s—1 generates the sesquialter ratio, and consequently the larger [number] from the computed ratio. **33** Therefore I believe that Cola Montanus interpreted this "sesquialtera" from "sesqui," which is *semis* [half] and *altera* [the other part], which is the middle, because the half of the other equal number is added to the whole number. **34** For one of two parts is called a half. **35** Others, however, suppose that "sesquialtera" is derived from "sesqui," which is the whole plus one part of two, that is, the middle, because the larger number in this [proportion] contains in itself the total smaller [number] and in addition half of this [number]. **36** I perceive that these [authorities] proceed more from the effect than from the meaning of the word itself. **37** Indeed, Pliny the Younger considers "sesqui" the whole with the

26–27 Ugol. *Decl.* 4.17.2; Boet. *De inst. mus.* 1.4.191.15–20; Boet. *De inst. arith.* 1.24.49.15–18 (cf. also 55); Nicom. *Intro. to Arith.* 1.19.1–2.

27a Ugol. *Decl.* 4.17.2; Boet. *De inst. arith.* 1.24.50.11–12; Nicom. *Intro to Arith.* 1.19.3.

27b Ugol. *Decl.* 4.18.2; Boet. *De inst. arith.* 1.24.50.3–12; Nicom. *Intro. to Arith.* 1.19.3. For a comprehensive discussion of this topic, see Kreyszig, *Franchino Gaffurio's "Theorica music"* Chapter 7, Section 4.

28 Ugol. *Decl.* 4.20.2; Nicom. *Intro. to Arith.* 1.19.4–5, 15.

29 Ugol. *Decl.* 4.20.2. This discussion of the species of the superparticular genus is continued in *Theorica* 3.4.8–13. The property of infinity of this genus is shown in the table provided in 3.4.13. Other properties of superparticular genera (i.e., the relationship between multiple and superparticular genera) are discussed in 3.4.30–48.

30 Ugol. *Decl.* 4.20.2.

33 Cf. also below 4.1.13a. This reference resists closer identification. On Cola Montanus, see: Mario Emilio Cosenza, *Biographical and Bibliographical Dictionary of the Italian Humanists and of the World of Classical Scholarship in Italy, 1300–1800,* 5 vols. and supplement (Boston: G. K. Hall, 1962–67), 3:2345.

37 Pliny jun. *Ep.* 4.9.9.

half, since he has written "sesquihora" [one and a half hour] for "an hour and a half." **38** Likewise, [Marcus Valerius] Martial has said the following in this poem: **39** "He for whom there is not yet a foot and a half penis." **40** Also, Varro in his books on agriculture [has mentioned] the "sesquimodius," that is, one and a half pecks. **41** Horace, too, in the *Art of Poetry* seems to have agreed with these writers in this poem: **42** "He throws forth swelling words and words that are a foot and a half long." **43** Nevertheless, it seems to me that the sesquialter does not agree at all with the superparticular proportions of this kind, **44** for if "sesqui" always means "whole plus the half," the sesquitertial and sesquiquartal as well as the remaining superparticular ratios, in which the larger number always comprises the smaller number with an aliquot part that is not the half, would not be able to be distinguished with their own names, **45** and for that reason they would have to be given other designations. **46** Saint Augustine in the first book of his *On Music* tries to clarify this "sesqui," as far as he can, in these words:

> **47** We give the term "sesqui" to two numbers when they are related to each other by such a ratio that the larger is larger than the smaller one by the same proportion by which it exceeds that number.

48 For, if the [ratio] 3:2 is made, the larger [number] exceeds the smaller by the third part of itself, **49** 4:3, by the fourth; **50** in the 5:6, by the fifth; **51** and thus, successively, the same proportion [applies] also to 6:4, 8:6, and 10:8. **52** From there it is easy to perceive and to explore this method in the corresponding larger numbers. **53** Moreover, I should not be able to give the origin of this name easily unless, perhaps, "sesqui" [is derived] from "se absque" or from "absque se" [literally, without itself], as for example, 5:4, as if 5 were a fifth larger "without itself" [i.e., without 4]. **54** So in Augustine the etymology of this word "sesque" is not absurd, even if perhaps it is not the one followed by [the theorists] who established this name. **55** Boethius in the first [book] of his *On Arithmetic* states that those numbers are considered "sesquati" [d6v] or

39 Mart. *Epigr.* 7.14.10.
40 Varro *Rust.* 1.42. Gaf. here confuses classical authors and sources. This passage is not found in Cato *Agr.* Such confusion presumably arises from the fact that the respective works of Varro and Cato were bound together in a volume that appeared in 1482 (cf. Heinrich Heill, ed., *M. Porci Catonis: De agri cultura liber; M. Terenti Varronis: Rerum rusticarum libri tres* [Leipzig: B. G. Teubner, 1884], Preface).
42 Hor. *Ars P.* line 97.
47–53 August. *De mus.* 1.10.17.
51 Note that these ratios of the multiple genus do not belong to the same species, that is, 6:4 represents the compound of the sesquialter proportion (3:2), 8:6 the compound of the sesquitertial proportion (4:3), and 10:8 the compound of the sesquiquartal proportion (5:4).
55 Ugol. *Decl.* 4.16.2, 6; Boet. *De inst. arith.* 1.24.49.15–17 (cf. above 26–27); Nicom. *Intro. to Arith.* 1.19.1.

superparticulars, whenever any larger number related to a smaller comprises this smaller number as a whole in itself, and some part of this smaller [number], **55a** for example, 4:3. **56** He calls this [proportion] sesquitertial, for 4 has in itself 3 once and one third part. **57** Indeed, superparticular ratios are formed whenever two equal numbers are related to each other in such a way that to one of them is added an aliquot part of its own quantity, which makes this number larger. **58** This latter number, when compared to the other unchanged number, will yield the "sesquated" proportion; **59** that is, the sesquated proportion will acquire its name from that part which has been superadded, as if we take two 4s and add to one of them a fourth, 5 results immediately, **60** and what was the fourth part of 4 now becomes the fifth of 5. **61** If we take this 5 to be compared with the first 4, we produce the sesquiquartal proportion, **62** and in that way the difference between the unchanged 4 and that which is augmented is 1, the added fourth part. **63** Consequently, 5 when it is in ratio with 4 is called "sesquiquartus," and the remaining numbers in like manner. **64** Boethius calls these larger "sesquati" numbers "leaders," the smaller "companions." **65** Accordingly, every proportion of this kind is named after the larger number, while keeping the smaller.

66 The superpartient genus occurs if a larger number compared with a smaller one contains this smaller number in itself only once **67** and in addition one non-aliquot part of this smaller number, **67a** which is formed from two or more aliquot parts, for example, 5:3. **68** For 5 contains in itself 3 only once and, besides, 2, which is not called an aliquot part of 3 **68a** but is composed of two aliquot parts of 3, that is, of two units, which are individual parts of 3. **69** The species of this genus are also infinite. **70** For, if that certain non-aliquot part of the smaller number by which the larger term surpasses this smaller one contains two aliquot parts of the smaller number itself, **70** the resulting proportion is called superbipartient, such as, 5:3, **70a** 10:6, and 15:9. **71** If,

55a–56 Ugol. *Decl.* 4.16.7 glossed.

57–63 Gaf.'s writing here reflects Boet. *De inst. arith.* 1.11.27.21–25; 1.14.30.15–28, 31.1–22.

64 Boet. *De inst. arith.* 1.24.49.27, 50.1.

66–67 Ugol. *Decl.* 4.34.2; Boet. *De inst. mus.* 1.4.191.23–25. Gaf.'s rather free rendition of the classical author includes a further expansion of the original statements. Nicom. *Intro. to Arith.* 1.20.1.

67a–68 Ugol. *Decl.* 4.34.3; Nicom. *Intro. to Arith.* 1.21.1.

69 The discussion of the species of the superpartient genus is continued in *Theorica* 3.4.14–23. The property of infinity of the genus is shown in the table provided in 3.4.23a.

70 Ugol. *Decl.* 4.34.6; 4.35.2; Boet. *De inst. mus.* 1.4.191.25–27; Boet. *De inst. arith.* 1.28.58.14–15.

70a The ratios 10:6 and 15:9 are compounds of the superbipartienttertial proportion (5:3).

71 Ugol. *Decl.* 4.36.2; Boet. *De inst. mus.* 1.4.191.27–28; Boet. *De inst. arith.* 1.28.58.16; Nicom. *Intro. to Arith.* 1.21.2. Here 8:5 is not a compound of 7:4; rather, 7:4 represents the

however, that part contains three aliquot parts of this smaller number, it is called a supertripartient proportion, **71a** for example, 7:4 **71b** and 8:5. **72** And if it has four parts, it is called a superquadripartient proportion, such as 9:5 **72a** and 11:7. **73** Therefore, the first species of this genus is the superbipartient proportion, **73a** the second the supertripartient, **73b** the third the superquadripartient, **74** and in this sequence one proceeds to infinity through a vigorous and diligent investigation.

75 Inasmuch as the species of this proportion have rather special names, [this] makes it necessary to investigate in which way those aliquot parts have in themselves another part with respect to the smaller term—whether these are thirds, fourths, or fifths. **76** If there are three parts, the proportion is called superpartienttertial; **77** if there are four parts, the proportion is called superpartientquartal, and so on. **78** Therefore let us proceed in this manner: **79** whenever a larger number contains an entire smaller number once and in addition a non-aliquot part, embracing two aliquot parts of the smaller number in itself, of which the individual aliquot parts are themselves thirds of this smaller number, then such a proportion is called [c1r] superbipartienttertial, such as 5:3. **80** If, in fact, a larger term contains a smaller term once and a certain part comprising three aliquot parts of the said smaller [number], **80a** of which each one of the three aliquot parts is a fourth part, this proportion is called supertripartientquartal, such as 7:4 **80b** and 14:8. **81** But if these three aliquot parts are fifth parts of this smaller number, the proportion is called supertripartientquintal such as 8:5 **81a** and 16:10. **82** And the process will be similar in other cases disposed according to a logical arrangement, **83** so that if the number 11 is related to the number 6, the proportion is called superquincupartientsextal. **84** Likewise, if the number 13 is related to the number 7, the proportion is called supersexcupartientseptimal. **85** And if 14 is related to 9, the proportion is called superquincupartientnonal, and so on.

---

supertripartientquartal proportion, whereas 8:5 expresses the supertripartientquintal proportion.

71a Ugol. *Decl.* 4.36.2; Boet. *De inst. mus.* 1.4.191.28; Boet. *De inst. arith.* 1.28.58.16; Nicom. *Intro. to Arith.* 1.21.1.

72 Ugol. *Decl.* 4.37.2 condensed; Boet. *De inst. arith.* 1.28.58.16–17. Here 11:7 is not a compound of 9:5; rather, 9:5 represents the superquadripartientquintal proportion, whereas 11:7 expresses the superquadripartientseptimal proportion. The next ratio in the series of superquadripartient proportions is 11:6 (see Nicom. *Intro. to Arith.* 1.21.1).

73–73b Ugol. *Decl.* 4.35.2, 36.2, 37.2; Boet. *De inst. arith.* 1.28.58.15–17; Boet. *De inst. mus.* 1.4.191.26; Nicom. *Intro. to Arith.* 1.21.2.

78–80 Gaf. employs ideas and concepts from Boet. *De inst. arith.* 1.28. Gaf. interprets the table of Boet. *De inst. arith.* 1.28.58.25–26; he also draws on Boet.'s terminology (e.g., "superpartiens tertias" as in *De inst. arith.* 1.28.60.4–9).

80a Boet. *De inst. arith.* 1.28.58.25–26 (cf. 78–80).

86 The multiple-superparticular genus is that in which a larger number related to a smaller comprises this several times in itself and in addition an aliquot part of this smaller number **86a** such as 5:2. **87** Therefore, if the larger number contains the smaller twice and in addition half of this smaller number, the proportion is called duple-sesquialter, such as 5:2, 10:4, **87a** and 15:6. **88** But if the larger number contains the smaller twice and in addition a third part of this smaller number, the proportion is called duple-sesquitertial, such as 7:3 **88a** and 14:6. **89** When the larger number contains the smaller number twice in itself and in addition the fourth part of this smaller number, the proportion is called duple-sesquiquartal, such as 9:4 and 18:8. **90** But when the larger number contains in itself the smaller three times and a half of this smaller number, the proportion will be triple-sesquialter, such as 7:2 and 14:4. **91** However, if in itself the larger number comprises the smaller three times and a third there will be a triple-sesquitertial proportion, such as 10:3 and 20:6. **92** And if the larger number contains the smaller three times and a fourth part of this smaller number, there will be a triple-sesquiquartal proportion, such as 13:4. **93** If the larger number contains in itself the smaller four times and a half of this smaller number, the proportion will be called quadruple-sesquialter, such as 9:2 and 18:4. **94** The diligent reader will perceive the remaining species of this genus more fully in our *Practica,* where we have arranged the measurable and singable proportions.

95 The multiple-superpartient genus is that in which a larger number related to a smaller contains this smaller number several times in itself and in addition one non-aliquot part of this, which contains several aliquot parts of this smaller, **95a** such as 8:3. **96** But if the larger term contains the whole smaller twice in itself and in addition a non-aliquot part, which has two aliquot parts of this smaller number, **96a** there will be a duple-superbipartient

86 Ugol. *Decl.* 4.38.3; Boet. *De inst. arith.* 1.29.61.1–3; Boet. *De inst. mus.* 1.4.191.29–30; 192.1–3. The genus is described without reference to terminology (cf. also 73b). Nicom. *Intro. to Arith.* 1.22.1.

86a Boet. *De inst. arith.* 1.29.61.30, 62.1; Nicom. *Intro. to Arith.* 1.22.3.

87 Ugol. *Decl.* 4.39.2, 4; Boet. *De inst. arith.* 1.29.62.17–18; Nicom. *Intro. to Arith.* 1.22.3.

87a Boet. *De inst. arith.* 1.29.62.17–18; Nicom. *Intro. to Arith.* 1.22.3.

88 Boet. *De inst. arith.* 1.29.63.1–2; Nicom. *Intro. to Arith.* 1.22.3.

90 Boet. *De inst. arith.* 1.29.63.17–18; Nicom. *Intro. to Arith.* 1.22.5.

91 Boet. *De inst. arith.* 1.29.64.1–2.

94 Gaf. *Pract.* 4.1–15.

95 Ugol. *Decl.* 4.47.2 glossed; Boet. *De inst. mus.* 1.4.192.11–14; Boet. *De inst. arith.* 1.31.65.1–6; Nicom. *Intro. to Arith.* 1.23.1.

95a Ugol. *Decl.* 4.47.2; Boet. *De inst. mus.*1.4.192.14–17; Boet. *De inst. arith.* 1.31.65.16–17; Nicom. *Intro. to Arith.* 1.23.3.

96 Ugol. *Decl.* 4.47.2, 6; Boet. *De inst. arith.* 1.31.65.16–17.

96a Ugol. *Decl.* 4.47.2, 6; Boet. *De inst. arith.* 1.31.65.16–17; Nicom. *Intro. to Arith.* 1.23.3.

proportion. **97** And if every single part of these aliquot parts is a third part of the smaller number, the proportion will be called **97a** duple-superbipartienttertial, **97b** such as 8:3. **98** However, if the larger number contains the smaller twice and in addition one certain part of this smaller number, enclosing in itself three fourths parts, the proportion is called **98a** duple-supertripartientquartal, **98b** such as 11:4. **99** If the larger number contains this smaller three times and in addition three fourths parts, there will be **99a** a triple-supertripartientquartal proportion, **99b** such as 15:4. **100** But if the larger number encloses the smaller number four times [e1v] and in addition one certain part made from three fifths parts of this smaller [number], the proportion will be a quadruple-supertripartientquintal, such as 23:5 and 46:10. **101** Thus a diligent investigator may obtain the remaining various species of this genus by joining the multiple and superpartient [genera] in proper aggregates and augmentations, always preserving their nature, as disclosed more fully in our *Practica,* in the discussion about the measurable proportions.

102 Likewise, it is necessary to know that the five genera of smaller inequality, as opposed to those previously mentioned, retain a similar progression and pattern, and also nature. **103** In this alone [the proportions] are known to differ, that the smaller numbers are contained in the genera of smaller inequality and precede the larger. **104** In the greater inequality, in fact, the larger numbers contain and precede the smaller numbers. **105** These are contrary operations and opposite effects in singable figures [*figurae cantabiles*], as is easily perceived in the *Practica*.

97a Ugol. *Decl.* 4.47.2, 6; Boet. *De inst. mus.* 1.4.192.16–17; Boet. *De inst. arith.* 1.31.65.12–13, 16–17 (cf. also above 1.1.95); Nicom. *Intro. to Arith.* 1.23.3.
97b The change is necessary to avoid repetition of an idea already stated in 95 of the 1480 edition. Boet. *De inst. mus.* 1.4.192.16–17; Boet. *De inst. arith.* 1.31.65.12–13, 16–17; Nicom. *Intro. to Arith.* 1.23.3.
98 Ugol. *Decl.* 4.48.2.
98a Ugol. *Decl.* 4.48.2, 5; Boet. *De inst. arith.* 1.31.65.13.
98b Ugol. *Decl.* 4.48.2, 5. Despite the slight difference in terminology, the authors are addressing the same proportion; Boet.'s "duplex supertripartiens" corresponds to Ugol.'s "dupla supertripartiens vel dupla superpartiens quartas." Gaf. employs an even more specific term, "dupla supertripartiens quartas," which both includes the components of the proportion and denotes their relation.
99 Ugol. *Decl.* 4.51.2.
99a Ugol. *Decl.* 4.51.2, 7; Boet. *De inst. arith.* 1.31.65.15.
99b Ugol. *Decl.* 4.51.2, 7. Boet.'s "triplex supertripartiens" corresponds to Ugol.'s "tripla supertripartiens vel tripla superpartiens quartas." As in the discussion of the preceding proportion, Gaf. employs a more specific term, "tripla supertripartients quartas."
100 Gaf. provides a logical extension of Ugol.'s and Boet.'s systematic discussion of proportions.
101 Gaf. here refers to the practical application of the proportions as set forth in composition and illustrated in his *Pract.* 4.
102–105 See Gaf. *Pract.* 4.2, 4, 6, 8, 10, 12.

106 Let us make a stricter formulation of the five previously mentioned genera in this manner. **107** Every larger term compared to a smaller contains either the smaller term itself several times and nothing beyond; then it is a multiple proportion. **108** Or it contains the smaller term itself once and something else besides, which is twofold: **109** [first], if that which the larger number itself contains beyond the smaller is an aliquot part of the smaller, the proportion is called superparticular; **110** [second] if that part which the larger number has beyond the smaller is not an aliquot part but a certain other part, comprising several aliquot parts of this smaller number, the proportion will be superpartient. **111** Or again, the larger number contains the smaller several times and something beyond, which is twofold; **112** [first], either that part which is beyond is the aliquot part of the smaller number, and then it is a multiple-superparticular proportion; **113** or [second], that which is beyond the smaller number is a non-aliquot part of the smaller number but a certain part which contains in itself several aliquot parts of this smaller number, and thus the proportion will be called multiple-superpartient. **114** A cautious reader can easily perceive the patterns of these genera in a nutshell from figure 3.3.1.

Chapter **Four**
The formation of the species of every possible genus

1 It is said that every single genus of inequality possesses infinitely many species; therefore, it is worthwhile to consider a brief examination of these species in the arrangement of the natural numbers. **2** After all numbers have been arranged in their natural order, we shall produce every species of the multiple genus whenever we join these individual numbers to unity. **3** Thus if 2 is related to unity, the first species of the multiple genus, that is, the duple proportion, will be formed. **4** If the number 3 is related to the same unity, the second species, that is, the triple proportion, will be produced. **5** If the number 4 is compared to the same unity, the third species, that is, the quadruple proportion, will be generated. **6** Moreover, the number 5, related to unity, generates the fourth species, that is, the quintuple proportion. **7** The progression is indefinite, as this table shows: [e2v]

106–114 This is a summary of *Theorica* 3.3. Figure 3.3.1 is reproduced in Boet. *De inst. arith.* 1.27.53.10–20 as well as in Nicom. *Intro. to Arith.* 1.19, 1.21.

1–7 Boet. *De inst. mus.* 2.4.229.11–24 paraphrased; Nicom. *Intro. to Arith.* 2.1.1–2; 2.3.1; 2.4.1. This is a summary of material already discussed in *Theorica* 3.3.15–24. The multiple genus is further examined in *Theorica* 3.5.2–12.

Figure 3.3.1

1	2	3	4	5	6	7	8	9	10
2	4	6	8	10	12	14	16	18	20
3	6	9	12	15	18	21	24	27	30
4	8	12	16	20	24	28	32	36	40
5	10	15	20	25	30	35	40	45	50
6	12	18	24	30	36	42	48	54	60
7	14	21	28	35	42	49	56	63	70
8	16	24	32	40	48	56	64	72	80
9	18	27	36	45	54	63	72	81	90
10	20	30	40	50	60	70	80	90	100

7a 2 3 4 5 6 7 8 9 10
1 1 1 1 1 1 1 1 1.

8 Out of number naturally arranged we shall try to enumerate all species of the superparticular genus. **9** If every number, except for unity, is in a ratio to a smaller number, its immediate neighbor, we may draw these species out to infinity in the right order. **10** Thus, if 3 is compared with 2, the first species—the sesquialter proportion—will result. **11** 4 related to 3 produces the second species—the sesquitertial proportion. **12** Out of 5:4 the third species occurs, that is, the sesquiquartal proportion, but 6:5 forms the fourth species, that is, the sesquiquintal proportion. **13** All species of the superparticular genus are known to proceed in this order, as this table makes apparent:

3 4 5 6 7 8 9 10
2 3 4 5 6 7 8 9.

14 If we wish to examine all species of the superpartient genus, we should

8–12 Boet. *De inst. mus.* 2.4.229.25–28, 230.1–7 paraphrased and glossed; Nicom. *Intro. to Arith.* 2.3.2.
8–13 This is a summary of material already presented in *Theorica* 3.3.26–30.
13 Boet. *De inst. mus.* 2.4.230.1–7.
14–18 Boet. *De inst. mus.* 2.4.230.8–12 paraphrased and glossed.
14–23 This is a summary of material already presented in *Theorica* 3.3.66–74. The superparticular genus is further examined in *Theorica* 3.5.30–48.

arrange the natural numbers not from unity and also not from the number 2, but rather from the number 3. **15** Indeed, if a larger number is in a ratio to this number 3, and one number lies naturally between both, the first species of the superpartient genus will be produced, such as 5:3, between which only 4 is placed. **16** And out of 5:3 the first species—the superbipartient proportion—is made. **17** The second species is produced from the naturally next smaller term, namely 4, which the larger term would exceed through the insertion of two natural numbers, that is, 5 and 6; **18** thus, 7:4 produces the second species—the supertripartient proportion. **19** The third species is proved to be accomplished from the next smaller number, that is, 5, and from the larger number related to it beyond the three naturally intervening numbers, for example, 9:5, from which the third species—the superquadripartient proportion—is made, **20** for the numbers 6, 7, and 8 fall naturally between 5 and 9. **21** If one progresses to the next smaller number in order, after the four numbers naturally falling between are passed over, the larger number then taken in relation to the smaller will form the fourth species of the superpartient genus—the superquincupartient proportion—such as 11:6, between which numbers fall naturally 7, 8, 9, and 10. **22** In like manner the remaining species may be explained, always observing, naturally, the increasing plurality of the intervening numbers. **23** This is clearly shown in these numbers:

23a 5 7 9 11 13 15 17 19
3 4 5 6 7 8 9 10.

24 Through the same order and process the diligent explorer may explain all species of the multiple-superparticular and of the multiple-superpartient genera, always associating the superparticular and the superpartient proportions with the multiple, as necessary.

25 It is well known, in light of what has been previously stated, that the multiple genus of inequality is by far more ancient, worthy, and fertile than the two remaining simple genera, because it is produced by comparing the numbers as they are naturally related to unity. **26** Now, the superparticular genus does not embrace unity, but rather those numbers which are imme-

19 Boet. *De inst. mus.* 2.4.230.8–12 paraphrased and glossed.
23a–24 Boet. *De inst. mus.* 2.4.230.13–19.
24 The multiple-superparticular as well as the multiple-superpartient genera are also discussed in *Theorica* 3.4.86–94, 95–101. The derivation of the multiple-superparticular ratios from multiple ratios is discussed in *Theorica* 3.5.58–61. The derivation of the multiple-superpartient ratios from superparticular ratios is examined in *Theorica* 3.5.62–68.
25 Cf. above 3.3.13.
25–30 Boet. *De inst. mus.* 2.5.230.21–28, 231.1-6.

diately next to unity; these are 3:2. **27** On the other hand, the superpartient genus seems to proceed by a very contrary means, since it couples not [e3r] successive but rather interrupted numbers and not always with an equal interruption, but now with 1, now 2, now 3, and even 4. **28** Thus the genus grows up to infinity. **29** Accordingly, the multiple genus grows out of unity, the superparticular from 2, and the superpartient from 3. **30** The genera seek their beginnings by means of the arrangement of the numbers. **31** For the multiple proportion possesses a greater force and power for musical consonance, after which comes the superparticular. **32** The superpartient genus is set apart, however, from the agreement of harmony as it seemed to Aristotle and the Pythagoreans, **33** although Ptolemy derives a single consonance, that is, the diapason-plus-diatessaron by means of a certain probable ratio from this genus. **34** For that reason we abandon this third genus and also combinations of the remaining two genera from the consideration of the consonances, **35** since through the first two simple genera, that is, the multiple and the superparticular, whose arrangement will be the subject of the following [chapter], we may produce and preserve the rationality of all melody.

Chapter **Five**
The production of the superparticular from the multiple [genera] and of the superpartient from the superparticular [genera]

1 If we wish to derive one genus from another and one species from another, as Boethius taught in his *On Arithmetic* and in his *On Music,* it is now and again necessary that the number approach the square. **2** That number is said to be square **2a** which grows equally in two dimensions, such as twice 2, three times 3, four times 4, five times 5, and so on. **3** A natural number is called the side of a square when it remains simple as it was before squaring. **4** Therefore, let us arrange the successive square numbers in natural order, and underneath them the natural numbers, excepting 1, which are the sides of these square numbers, **4a** in this order:

4 9 16 25 36 49 64 81 100
2 3 4 5 6 7 8 9 10.

5 Note that I said that the square numbers are by nature successive, because between these numbers no other square number is found. **6** Therefore, in this

33 Cf. above 2.5.40ff.
2–2a Boet. *De inst. mus.* 2.6.231.13–14.
4a–8 Boet. *De inst. mus.* 2.6.231.18–22 paraphrased and glossed.

figure, the upper ones are the square numbers or squares. **7** To be sure, the lower numbers are the sides of the squares. **8** Therefore, if we compare each and every square with its side, the multiple ratios will be produced; **9** for the first square, that is, the 4, related to its side, forms the first multiple, that is, the duple proportion. **10** The second square, that is, the 9, compared to its side, effects the second multiple, that is, the triple proportion. **11** But if you compare the third square, that is, the 16, to its side, the third species of the multiple genus, that is, the quadruple proportion, results. **12** Thus are produced the species of the multiple genus.

13 Let us reveal a certain speculative consideration concerning these squares. **14** If we remove successively smaller from successively larger squares, that part of the larger square which remains will be as large as one side of each square joined together, **15** so that if I remove 4 from 9, 5 will remain. **16** Thus [e2v], if 2 and 3, which are the sides of the two squares, are joined, they yield 5. **17** In the same manner, if we subtract 9 from 16, 7 will remain, which is joined from 3 and 4, that is, from the sides of either square. **18** This also happens with the two sides of any square. **19** If the squares are not successive, but one intervenes in the natural order, then the sides of these squares joined will complete only the half of the remaining part of the larger square; **20** so if we remove 4 from 16, 12 remains, whose half is 6, which is coupled from the sides of both squares; **21** these sides are 2 and 4. **22** The same applies to other squares, **22a** as long as one square number always naturally falls between. **23** But if two square numbers naturally intervene, then that which remains from the larger square after the withdrawal of the smaller square, multiplied three times, equals the sides of these squares taken together; **24** or the sides added together will equal the third part of the remainder, **24a** so that if we remove 4 from 25, 21 remains. **25** In fact, the sides of these squares are 2 and 5, which, when joined, total 7, and 7 is a third of 21. **26** The rule, then, is: **27** if three square numbers naturally intervene, the sides of these squares joined will equal the fourth part of that which remains after the smaller square has been removed from the larger square. **28** If, further, four square numbers naturally fall between, the sides of these will produce only a fifth part of this square's remainder. **29** Thus the parts have been diminished by one number more than there are intervening square numbers.

30 Since successive square numbers have naturally been shown to produce

8 Continuation of the discussion of multiple proportions in *Theorica* 3.3.13–25, 3.4.2–7.
9–13 Boet. *De inst. mus.* 2.6.231.18–21 paraphrased and glossed.
13-29 All this follows from $a^2 - b^2 = (a + b)(a - b)$.
9–13 Boet. *De inst. mus.* 2.6.231.18–21 paraphrased and glossed.
14–22 Boet. *De inst. mus.* 2.6.231.21–28, 232.1–10.
23–24 Boet. *De inst. mus.* 2.6.232.1–12 paraphrased.
24a–29 Boet. *De inst. mus.* 2.6.232.12–20.
30–31, 31b Boet. *De inst. mus.* 2.7.232.23–26 paraphrased.

multiple ratios which preserve the arrangement of a square's sides, it remains to elucidate more clearly how inequality is derived from equality in the following manner. **31** Just as arithmeticians wish that all numbers be congenitally produced from a single unity, so it is established, **31a** according to the doctrine of Boethius, **31b** that superparticularity results from multiplicity, and the superpartient genus from the superparticular, and the other genera in the same manner. **32** From reversed multiple species are derived superparticular proportions; **33** and from transformed superparticular proportions are generated the superpartient. **34** Now let three 1s be arranged in a series and three numbers laid under them, **35** the first [lower] number being equal to the first unity, the second, to the first [unity] plus the second, the third, [equal] to the first [unity] plus twice the second plus the third. **36** This twice yields the first species of the multiple genus, that is, the duple proportion, as here:

1 1 1
1 2 4.

37 But if we place the numbers 2 or 3 in the upper series, and underneath them we place corresponding numbers, we have very little doubt that duple proportions will result. **38** Now if two of the newly created duple numbers are placed in the upper series, and below these are placed the corresponding values, the triple ratios will arise, in this way:

1 2 4
1 3 9.

39 But if we arrange these triple numbers in the upper series, we shall obtain quadruple ratios by arriving at the lower series through the same process as before, namely:

1 3 9
1 4 16.

40 Other multiple ratios will be obtained in turn by the same process [e4r]. **41** Thus, the triple ratios arise from the duple, the quadruple from the triple, **41a** the quintuple from the quadruple, **41b** and the others are accustomed to be

32–36 Boet. *De inst. mus.* 2.7.232.26–31; Boet. *De inst. arith.* 1.32.67.24–27, 68.1–2; Nicom. *Intro. to Arith.* 1.23.9–10.
37 Boet. *De inst. mus.* 2.7.233.7–8.
38 Boet. *De inst. mus.* 2.7.233.8–11; Boet. *De inst. arith.* 1.32.68.3–11; Nicom. *Intro. to Arith.* 1.23.9.
39 Boet. *De inst. arith.* 1.32.68.12–20 paraphrased; Nicom. *Intro. to Arith.* 1.23.9.
40–41 Boet. *De inst. mus.* 2.7.233.12–13; Boet. *De inst. arith.* 1.32.68.21–24, 69.1.
41a Boet. *De inst. mus.* 2.7.233.13.
41b Boet. *De inst. mus.* 2.7.233.13–14; Boet. *De inst. arith.* 1.32.69.2–4; Nicom. *Intro. to Arith.* 1.23.9.

generated in the same manner. **42** Now, in like manner we derive the sesquialter proportions from the duple proportions as follows. **43** After the [order of the] numbers of the duple [proportions] has been reversed, so that the larger are first, [and they are placed in the upper series] in this arrangement, **44** the numbers [of the lower series] are put underneath in the way already proposed for the formation of the multiple [genera]: the first [lower] number is equal to the first [upper]; the second, to the first plus the second [upper]; the third to the first plus twice the second plus the third:

4 2 1
4 6 9.

45 Through the same process, sesquitertial proportions may also be generated, provided that the triple proportions have been properly arranged and reversed, as here:

9 3 1
9 12 16.

46 If now quadruple numbers which have been reversed are arranged, the sesquiquartal ratios will be produced (all errors having been removed) as here:

16 4 1
16 20 25.

47 Likewise sesquiquintal proportions are produced from the quintuple numbers that have been reversed. **48** And in the same manner, with similar names given to either part, superparticularity is generated ad infinitum from multiplicity.

49 To be sure, from superparticular proportions which have been transformed and which proceed through the same arrangement as the preceding, superpartient proportions arise. **50** The numbers of the sesquialter proportions are arranged in this order: 9, 6, 4. **51** Let the first [lower series number] be equal to the first [upper number], that is, 9; the second [equal to] the first plus the second [upper], that is, 15; the third [equal to] the first plus twice the second plus the third, namely 25. **52** All these numbers are detailed in this table:

42–44 Boet. *De inst. mus.* 2.7.233.14–23; Boet. *De inst. arith.* 1.32.69.5–9; Nicom. *Intro. to Arith.* 1.23.9.
45 Boet. *De inst. arith.* 1.32.69.10–19 condensed; Nicom. *Intro. to Arith.* 1.23.9. The content of this passage is also hinted at in Boet. *De inst. mus.* 2.7.233.24.
46 Boet. *De inst. arith.* 1.32.69.20–25 paraphrased; Nicom. *Intro. to Arith.* 1.23.9. The content of this passage is also hinted at in Boet. *De inst. mus.* 2.7.233.24–25.
48–52 Boet. *De inst. mus.* 2.7.233.25–29, 234.1–5; Boet. *De inst. arith.* 1.32.70.1–16; Nicom. *Intro. to Arith.* 1.23.9.

9 6 4
9 15 25.

53 Accordingly, the superpartient proportion is produced from the superparticulars which have been reversed in this manner, **54** and the superbipartients are made from the sesquialter proportions which have been reversed. **55** If you consider this discussion diligently, you discern that the supertripartient proportions are obtained from the sesquitertial proportions which have been reversed, **55a** as here:

16 12 9
16 28 49.

56 And from the sesquitertial proportions which have been reversed, superquadripartient proportions are produced, **57** and the careful investigator derives the other proportions by similar designations from the superparticulars which have been reversed.

58 The multiple-superparticular ratios are generated not from superparticular [proportions] that have been reversed but are produced from multiple proportions which maintain their numerical order, **58a** as shown here:

4 6 9
4 10 25.

59 Thus the duple-sesquialter ratios are produced from the sesquialter, **60** and the duple-sesquitertial from the sesquitertial, as here:

9 12 16
9 21 49.

61 In the same manner the remaining duple-sesquiquartal ratios [proceed] from the sesquiquartal, and so on.

62 Surely, from unconverted superpartients, **62a** as they are produced from the superparticulars, the multiple-superpartient proportions **62b** are assembled in the same order, **62c** as here:

53 Nicom. *Intro. to Arith.* 1.23.10.
54–55 Boet. *De inst. mus.* 2.7.234.6–9.
55a–56 Boet. *De inst. arith.* 1.32.70.17–25, 71.1–4 condensed; Nicom. *Intro. to Arith.* 1.23.10.
57–58 Boet. *De inst. mus.* 2.7.234.9–14.
58a Boet. *De inst. arith.* 1.32.71.5–17 condensed; Nicom. *Intro. to Arith.* 1.23.11.
59–60 Boet. *De inst. arith.* 1.32.71.17–23; Nicom. *Intro. to Arith.* 1.23.11.
62 Boet. *De inst. mus.* 2.7.234.14.
62a Boet. *De inst. arith.* 1.32.71.24–27 condensed; Nicom. *Intro. to Arith.* 1.23.14.
62b–62c Boet. *De inst. arith.* 1.32.71.27–28, 72.1–5; Nicom. *Intro. to Arith.* 1.23.14.

9 15 25
9 24 64.

63 Then the duple-superbipartients are produced from the superbipartients, **64** and if the supertripartients are extended in the same way, they will form one duple-supertripartient and another proportion of the same genus, but not of the same species, as is shown in this figure:

4 7 10
4 11 28.

65 For 11:4 is a duple-supertripartient proportion, while 28:11 is a duple-supersexcupartient proportion, **66** for it contains this [i.e., 11] twice and in addition, six parts of it, that is, six units. **67** And in the same manner let us arrange the other supertripartients proportions in this order:

5 8 11
15 13 32.

68 Indeed, 13:5 yields the duple-supertripartient proportion, but 32:13 forms the duple-supersexcupartient proportion. **69** Thus from superpartient proportions are assembled the multiple-superpartient proportions [e4v], whatever these may be. **70** This is evident to the more diligent observer. **71** Thus it is established that every form of inequality is produced from equality under the influence of nature [by natural law]. **72** First, the duple proportions arise from three equal terms which have been arranged; **73** then the triple, thereafter the quadruple, and the following ones in the same order. **74** Again, the superparticular proportions are assembled from the reversed multiple which have been arranged; **75** the sesquialter from the duple, the sesquitertial from the triple, the sesquiquartal from the quadruple, and similarly the remaining. **76** On the other hand, we see that the superpartients are produced from superparticular proportions which have been reversed, **77** so that the sesquialter forms the superbipartient, the sesquitertial the supertripartient, and the sesquiquartal the superquadripartient, and the remaining in the same manner.

78 The multiple-superparticular proportions are derived from the previous superparticulars, which are in the right order and not reversed. **79** And the multiple-superpartient proportions are derived directly from the superpartients. **80** Consequently, these three rules have been written down **80a** by Boethius in the last [chapter] of the first book of his *On Arithmetic* for the derivation of inequality from equality: **81** first, we place the first number as equal to the first number; **82** second, the second number, equal to the first plus

63 Boet. *De inst. arith.* 1.32.72.6–7 glossed; Nicom. *Intro. to Arith.* 1.23.14.
71–80, 81–85 Boet. *De inst. arith.* 1.32.67.1–20.

the second; **83** third, we make the third number equal to the first plus twice the second plus the third. **84** Because if we have perfected the proportions in equal terms, duple proportions will be produced, from which triple proportions arise by the same method. **85** And also quadruple proportions arise from triple, **85a** and quintuple from quadruple, **85b** as here:

1 1 1
1 2 4
1 3 9
1 4 16
1 5 25.

86 A certain secret arrangement of nature that is well known from the explanation of our Boethius in the first [chapter] of the second [book] of his *On Arithmetic,* remains worth knowing. **86a** Just as all the species of inequality originate from the margin of equality, as we have learned, **87** so all inequality resolves again into equality, as if to a certain element of its own genus. **88** This element can be gathered together from a threefold order; for example, if any three terms are arranged, as long as they are unequal terms and they are arranged proportionately, **89** so that the middle term holds the same ratio to the first as the last to the middle, and this is any ratio of inequality, whether multiple, superparticular, or superpartient, **90** or, if the ratio is in one composed from these genera of proportions, one and the same ratio will without doubt be exhibited. **91** To be sure, after the three terms have been arranged in equal proportions, as stated, we always subtract the last term from the middle, **92** and we rearrange this last term as the first term, **92a** and the remainder from the middle term as the second. **93** From the third sum of the numbers which have been put forth, we subtract one first number and two second numbers—those which remain from the middle—**94** and what remains of the third sum we shall set up as the third term. **95** When this is done, the sums of the terms revert to a smaller size, **96** and the proportions are reduced to the more primary ratios, **97** just as if the quadruple proportion is at first related to the triple, thereafter to the duple, and then to equality, there will always remain that primary ratio. **98** If we started with a superparticular proportion, however, the sesquiquartal will first be reduced to the sesquitertial, then to the sesquialter, and finally to three equal terms. **99** And since this

85b Boet. *De inst. arith.* 1.32.68.16–20; this part is an extrapolation of a longer passage (*De inst. arith.* 1.32.67.24–27, 68.1–20) which Gaf. already quoted in *Theorica* 3.5.32–39. The last line in this table ("1 5 25"), not present in Boet., represents a logical extension of the system; it also refers to *Theorica* 3.5.85a.
86a–92 Boet. *De inst. arith.* 2.1.77.15–19, 78.1–11.
92a–94 Boet. *De inst. arith.* 2.1.78.11–15 quoted verbatim.
95–124 Boet. *De inst. arith.* 2.1.78.16–31, 79.1–28.

example has been derived from the remaining proportions, let us show that it occurs only in the multiple proportion. **100** The same method or rule will be evident to someone who is diligent in experimenting also with the other species of inequality [e5r].

101 Let us establish three quadruple terms in this way: 8, 32, 128. **102** When I remove the smaller term from the middle, that is, 8 from 32, 24 will remain. **103** Let us assume 8 as the first number and, as the second term, the remainder from the middle, that is, 24. **104** Let these two terms, then, be 8 and 24. **105** From the third term, that is, 128, let us subtract once the first term, that is, 8, and twice the second, that is, two times 24; **106** 72 remains. **107** After these terms are arranged, however, a triple proportion closer to equality is reduced from the quadruple. **108** These numbers are 8, 24, 72. **109** But if we likewise make a proportion from these terms, the relation will go back to the duple proportion.

110 Assume a first term equal to the smaller term, that is, 8, **111** and let the first term be subtracted from the second; **112** 16 will remain. **113** Then from the third term, that is, 72, let us remove the first, that is, 8, and twice the second, that is, twice 16; **114** the remainder will be 32. **115** If these terms are arranged, the ratios of duple proportions result, as here: 8, 16, 32. **116** But if we likewise make a proportion from these terms, we shall perceive that all these terms are reduced to the sums of equality.

117 Let us arrange a first term equal to the smaller term, that is, 8, and let us subtract it from 16; **118** 8 will remain. **119** After these have been arranged, if we remove the first, that is, 8, from the third term, that is, 32, and twice the second term, that is, 8, 8 remains—**120** which numbers thus arranged, the first equality emerges in this order: 8, 8, 8.

121 Anyone who directs his thoughts to other species of inequality will without doubt find the same correspondence, **122** for which reason one must not at all doubt that inequality is the mother of the quantity which is related to anything, **123** as unity is called the beginning and also the first element of quantity, constant per se. **124** Also, inequality is generated from equality, and its final dissolution is made in this equality.

Chapter **Six**
The investigation of more similar superparticular proportions

1 Boethius shows that from multiplicity many superparticular proportions, and, consequently, as many as one wants are derived directly by this

1 Boet. *De inst. mus.* 2.8.234.19–25 condensed.

rule, after absolutely every error has been removed. **2** Every single multiple proportion, that is, reckoned from unity, precedes as many superparticular proportions in the opposite part of its denominator as this multiple proportion recedes from unity in this fashion: **3** the duple proportion precedes the sesquialter, the triple the sesquitertial, the quadruple the sesquiquartal, and so on. **4** Therefore, let us combine several **4a** duple ratios **4b** from unity with its successive sesquialter proportions **4c** in this order, **4d**

1	2	3	8	16
	3	6	12	24
		9	18	36
			27	54
				81.

5 In this disposition of the duple numbers, the multiple proportion involving 2 has only one such number, that is, 3, which forms the sesquialter with it, because the first number separates from unity. **6** But since this 3 cannot be divided into two equal parts, it receives no one number which to itself can maintain the sesquialter proportion. **7** Surely 4, which is the second duple number counting from unity, is shown to precede two sesquialter proportions, assembled from the number 6 to itself [i.e., to the 3] and from the number 9 to the number 6 [e5v]. **8** To be sure, the number 9 is not considered from any sesquialter number, inasmuch as it cannot be divided by a true half. **9** The same order is present with regard to the remaining numbers. **10** Also it is evident that the triple numbers create the sesquitertial in the same progression, **10a** as that representation shows:

1	3	9	27	81
	4	12	36	108
		16	48	144
			64	192
				256

11 In fact, this arrangement of the triple numbers shows that when one starts from unity, 3 is the first triple number and is the only antecedent number that forms a sesquitertial ratio **11a** with 4. **12** Since 4 does not have a third part, it

2–3, 4a, 4c Boet. *De inst. mus.* 2.8.234.26–30, 235.1.
4d Boet. *De inst. mus.* 2.8.235.3–7. Nicom. *Intro. to Arith.* A slight expansion of this table is also found in Boet. *De inst. arith.* 2.2.81.22–26.
10a Boet. *De inst. mus.* 2.8.235.16–21. A slight expansion of this table is also found in Boet. *De inst. arith.* 2.2.82.21–27.
11 Boet. *De inst. mus.* 2.8.235.22–24. The substance of this passage is also found in Boet. *De inst. arith.* 2.2.81.28, 82.1–2.
11a–12 Boet. *De inst. arith.* 2.2.82.3–9 paraphrased; Nicom. *Intro. to Arith.* 2.3.4.

contains no number which may relate to this number in a sesquitertial ratio. **13** On the other hand, 9, because it is the second disjunct triple number counting from unity, precedes two sesquitertial numbers, that is, 12 and 16. **14** But since 16 does not possess a third part, no number can be related to it in the sesquitertial ratio. **15** Likewise, given that 27 is the third triple number from unity, it is necessary that it precede only three sesquitertial numbers. **16** And so the third part is always confined by a certain limit in the most remote sesquitertial number so that no sesquitertial can have more numbers. **17** But if I arrange the aggregate of the quadruple proportions, we would produce the sesquiquartal proportions by the same rule. **18** If I arrange the quintuple proportions, we would produce the sesquiquintal proportions, and so on. **19** The individual numbers of the multiple denomination precede as many superparticular proportions as the degree they depart from unity itself. **20** This theorem has been invented in such manner that however many sesquialter, sesquitertial, or other superparticular ratios a person tried to assemble, he would never be subject to error. **21** Now also, there remains in this discussion to derive the reason why the roots of proportions in any genus are named by those numbers which first naturally form those proportions. **22** Therefore, every first proportion with its terms in the order of natural numeration **22a** are roots of all similar proportions, as for example, 2 to unity; **23** since this is the first duple proportion, it is said to be the root of all duple proportions. **24** And 3:2 is the first sesquialter proportion in natural numeration. **25** These numbers and this proportion produce the roots of all sesquialter proportions. **26** Similarly, we call the sesquitertial proportion formed from 4:3 **26a** the root of the remaining sesquitertial proportions. **27** And the remaining proportions are formed in the same manner. **28** Now, similarly, let us investigate two adjacent sesquialter proportions in another progression. **29** After the sesquialter root has been taken, that is, 2 and 3, then 2 multiplied by 2 yields 4; **30** 3 multiplied by 2 yields 6; **31** and 3 multiplied by itself yields 9. **32** These numbers may be arranged in the following order:

2 3 3
4 6 9.

33 Consequently, two sesquialter proportions have been produced in this distribution, that is, 6 to 4, and 9 to 6. **34** But if we tried to derive three

13–15 Boet. *De inst. arith.* 2.2.82.3–9 paraphrased; Nicom. *Intro. to Arith.* 2.4.1.
16–19 Boet. *De inst. mus.* 2.8.235.24–29.
20 Boet. *De inst. mus.* 2.8.236.6–10.
22a–25 Boet. *De inst. mus.* 2.8.236.22–25 paraphrased and glossed.
26, 28–41 Boet. *De inst. mus.* 2.8.236.25–29, 237.1–25.

[proportions], we would arrange the same numbers—which have been disposed in investigating the two numbers—and also these two sesquialter proportions, which have been investigated. **35** Let us multiply 4 by 2, which yields 8; **36** 6 by 2 yields 12; **37** 9 by 2 yields 18, **38** and 9 by 3 yields 27, in this arrangement:

2 3 4 6 9 9
8 12 18 27.

39 With the remaining numbers the same process will obtain.

40 To investigate the sesquitertial proportions, let us arrange their roots, that is, 4 and 3, and assemble these numbers by an equivalent multiplication. **41** But should we seek the sesquiquartal proportions, let us arrange their roots, and by the same multiplication any sesquiquartal will be extended. **42** Similarly, we shall easily assemble the remaining proportions [e6r] in the same order. **43** In addition, one may consider the following: **44** if the difference of two numbers measures them entirely, these numbers, which the difference has measured, will be in the same proportion. **45** In this proportion there will also be these numbers, according to which their difference has measured them, as for example, let us take 50 and 55. **46** They are related to one another in the sesquidecimal ratio. **47** For the difference between them is 5, which is the tenth part of 50; **48** for 10 taken five times measures 50. **49** Of 55, however, the eleventh part is 5, **49a** since it received the number eleven times, traverses 55 in measuring. **50** Therefore the proper difference, that is, 5, measures out the numbers 50 and 55, ten and eleven times, respectively. **51** In fact 11 and 10 are related in the sesquidecimal proportion. **52** So it is established that the numbers which their proper difference has measured completely **52a** are in this proportion **52b** in which there are these numbers which the proper difference has measured out. **53** If any particular difference in numbers measures those numbers, however, so that the plurality exceeds the same measure of the numbers **54** and if there also is the same excess in both, and if the measure of the difference is smaller than the plurality of the numbers, **55** then the numbers—if that which remains of these numbers after measuring has been subtracted—will obtain alternately to themselves a larger proportion than were the integers when the proper difference measured them. **56** For instance, let there be two numbers, 53 and 58. **57** Accordingly, 5, which is their difference, measures them. **58** Certainly 5 measure 53 ten times up to 50, and 3 remains. **59** Again, the same 5 measures the 58 eleven times up to 55,

44–49, 50–52, 52b–60 Boet. *De inst. mus.* 2.9.237.29, 238.1–27.

and again 3 remains. **60** Therefore, 3 taken away from both will leave 50 and 55. **61** It is evident that a larger proportion lies between 50 and 55 than between 53 and 58 **62** because the larger proportion is always perceived in smaller numbers and the smaller proportion in larger numbers. **63** If, however, that measuring-out of the difference surmounts the multitude of the numbers and the same measuring-out omits either of the numbers in the plurality, **64** the proportions will be smaller numbers measured above with the addition of its sum by which the difference measuring both numbers surmounts them than they were before their proper difference measured them. **65** Take the numbers 48 and 53, whose difference is 5, **66** and let 5 measure the number 48. **67** Taken ten times, it forms 50. **68** Consequently 50 exceeds 48 by 2. **69** Moreover, the same 5 measures 55 eleven times; **70** then 55 again exceeds the number 53 by the same 2. **71** Let 2 be added to both numbers, **72** and let them be arranged in their order: 48, 53, 50, 55. **73** The smaller proportion is 50:55, resulting from the addition of the number 2, by means of which the difference surmounts the numbers 48 and 53. **74** This difference of 5, which rises beyond only in these numbers, measures them out. **75** Certainly the larger and the smaller proportions are understood in this manner, as the half part is larger than the third part, **76** the third part is larger than the fourth part, **77** and, the fourth part is larger than the fifth part, and so forth. **78** Thus, the sesquialter proportion is larger than the sesquitertial, and [e6v] the sesquitertial exceeds the sesquiquartal, **79** and also the sesquiquartal exceeds the sesquiquintal proportion, and so for the remaining. **80** Hence it happens that among smaller numbers, the larger proportion is always the superparticular ratio, as is proved in this natural numerical arrangement: 1, 2, 3, 4, 5. **81** The sesquialter is 3:2; **82** 4:3, the sesquitertial; **83** and 5:4, the sesquiquartal. **84** The larger numbers, in fact, are 4 and 5; **85** the smaller, 3 and 2. **86** Also, the proportion 3:2 is larger than 4:3. **87** Likewise, it is larger than 5:4. **88** From this it is also shown that if an equal plurality is added to the terms of a superparticular proportion, the proportion will always be larger before the addition of the plurality than after.

61–62 Boet. *De inst. mus.* 2.9.238.31–32, 239.1.
63–65 Boet. *De inst. mus.* 2.9.239.2–8 quoted verbatim.
66–80 Boet. *De inst. mus.* 2.9.239.8–28.
81–82 Boet. *De inst. mus.* 2.9.239.29, 240.1. Gaf. omits Boet.'s initial statement concerning the numerical series given in *Theorica* 3.5.80 "Binarius igitur ad unitatem duplus est."
84 Boet. *De inst. mus.* 2.9.240.1–2. Gaf. here deliberately substitutes Boet.'s "ternarius et quaternarius" for "quaternarius et quinarius," thus providing a connection to the preceding statement (*Theorica* 3.5.83) absent in Boet.'s text.
85 Boet. *De inst. mus.* 2.9.240.2–3. Gaf.'s truncation of Boet.'s statement ("minores ternarius et binarius et unitas") is consistent with his change of Boet.'s text in *Theorica* 3.5.80.
88 Boet. *De inst. mus.* 2.9.240.4–8.

Chapter **Seven**
The threefold proportionality

1 Whenever three or more successive terms are arranged to form two or more successive proportions, this is called proportionality. **2** Mathematicians call this relation of two or more proportions a "conjunction." **3** They are accustomed to distinguish three kinds, namely, the arithmetic, the geometric, and the harmonic. **4** When three terms only are established, and the first term to the second retains the same proportion as the second to the third, then there is said to be proportionality, for the second is the middle term of the three. **5** Indeed, the proportions of the extreme terms and of the middle terms exhibit a threefold participation. **6** Or the difference between the smaller term and the middle, and between the middle and the largest may be equal, **7** but the proportion in this case is not equal, so that in these numbers, 1, 2, 3 **7a** unity constitutes the difference. **8** Nevertheless the proportion is not equal, **9** for 2:1 constitutes the duple proportion, whereas 3:2, the sesquialter. **10** But a proportion that is equal with either term but does not have equal differences is one such as formed with these numbers, 1, 2, 4: **11** 2:1 forms the duple proportion, as does 4:2. **12** But between the 4 and 2, the difference is 2, whereas between 1 and 2, it is 1. **13** Or the equality may rest in neither the same proportions nor in the same differences, **14** but as the larger term relates to the smallest term **14a** or the extreme to the extreme, **14b** so the difference between the larger terms relates to the difference between the smaller terms, as in these numbers 3, 4, 6. **15** For 6:3 yields the duple proportion, **16** and between 6 and 4, the difference is 2. **17** But between 4 and 3, the difference is 1. **18** Thus between 2 and 1 the duple proportion is made, just as between the extremes, **18a** that is, 6:3. **19** Therefore Boethius and Albertus [Magnus] and

1 Boet. *De inst. arith.* 2.40.137.8–10; Boet. *De inst. mus.* 2.12.241.16–20.

3 Boet. *De inst. arith.* 2.41.139.2–3; Nicom. *Intro. to Arith.* 2.21; 2.24; 2.25; 2.27. Boet. here uses the term "medietas" for the more common "proportionalitas" (cf. Boet.'s chapter heading).

4 Boet. *De inst. mus.* 2.12.241.20–23; Boet. *De inst. arith.* 2.43.141.3–7 paraphrased.

5 Boet. *De inst. mus.* 2.12.241.23–24.

6–7 Boet. *De inst. mus.* 2.12.241.24–27 quoted verbatim.

7a–8 Boet. *De inst. mus.* 2.12.241.27–28, 242.1; Boet. *De inst. arith.* 2.43.141.9–11; see also Gaf. *De harm.* 3.1.

9 Boet. *De inst. mus.* 2.12.242.1; Boet. *De inst. arith.* 2.43.141.11–12.

10 Boet. *De inst. mus.* 2.12.242.2–3 quoted verbatim; see also Gaf. *De harm.* 3.3.

11–12 Boet. *De inst. mus.* 2.12.242.3–7; Boet. *De inst. arith.* 2.43.141.14–17.

13–14, 14b–18 Boet. *De inst. mus.* 2.12.242.7–17; see also Gaf. *De harm.* 3.5.

19 Al. Mag. *De quant.* 1.195.32; see also A. G. Molland, "Mathematics in the Thought of Albertus Magnus," in *Albertus Magnus and the Sciences: Commemorative Essays 1980,* ed. James A. Weisheipl (Toronto: Pontifical Institute of Mediaeval Studies, 1980), pp. 463–478. The "other mathematicians" are presumably Hippasus, Philolaus, and Archytas. Cf. Hermann

the other mathematicians **19a** call arithmetic proportionality that in which the differences are equal, **19b** yet the proportions are unequal [f1r]; **20** they call geometric proportionality that in which the proportions are equal but the differences are unequal; **21** and they call harmonic that proportionality in which the proportion of the extreme terms is equal to the proportion of the differences. **22** Let us arrange these proportionalities in the following formulas [given in figures 3.7.1, 3.7.2, and 3.7.3].

23 Therefore, the proportionality of these three means is properly and rightly called geometric, because it is made up of totally equal proportions. **24** Nevertheless, we use the same designation without distinction in naming the other proportionalities. **25** Certainly, of these, the one proportionality is continuous, the other is disjunct. **26** A proportionality is continuous when one number alone is placed as the middle, **27** which then is put sometimes in comparison after the smaller [number]; sometimes it is put before the larger [number], **27a** as appears in the figures represented above. **28** A proportionality is disjunct when two terms are placed as the middle, as in the geometric [proportionality] in this manner; 1, 2, 3, 6. **29** For as the 2 is to 1 so is 6 to 3, and this proportionality is called disjunct. **30** Hence, one is permitted to understand that the continuous proportionality consists of three terms, a disjunct [proportionality], in four **30a** or several. **31** So, in the second [book] of his *On Arithmetic,* Boethius asserts that the arithmetic proportionality conforms to a government which is ruled by a few for the reason that its smaller terms have the larger proportion. **32** In fact, [Boethius declares], the geometric [proportionality] is somehow a popular equality of a city, because it is made up, whether of majorities or minorities, through an equal proportionality of all [terms]. **33** And among all the terms, there is a certain parity of the middle, which conserves an equal right in the proportions. **34** Also Boethius declares that the harmonic proportionality calls to mind the government of the nobility, because the larger proportionality is found in the larger terms. **35** But I did not care to arrange these ten proportionalities, which

---

Diels, *Die Fragmente der Vorsokratiker,* 1:110 (Hippasus 18.15); 1:404–405 (Philolaus 44.A24); 1:435–436 (Archytas 47.B2).

19a–22 Boet. *De inst. mus.* 2.12.242.17–23.

23–24 Boet. *De inst. mus.* 2.12.242.26–31.

25–27, 28–30 Boet. *De inst. mus.* 2.13.243.2–5.

30a With these words in the *Theorica,* Gaf. indicates a continuation of Boet.'s text, which he does not quote in the subsequent passage.

31 Boet. *De inst. arith.* 2.45.149.7–9; Gaf. *De harm.* 3.8.

32–33 Boet. *De inst. arith.* 2.45.149.11–16; Gaf. *De harm.* 3.8.

34 Boet. *De inst. arith.* 2.45.149.9–11; Gaf. *De harm.* 3.8. In the *Theorica* Gaf. has substituted "harmoniam" for Boet.'s "melodiam."

35 Reference to Boet. *De inst. arith.* 2.40–44.

Figure 3.7.1

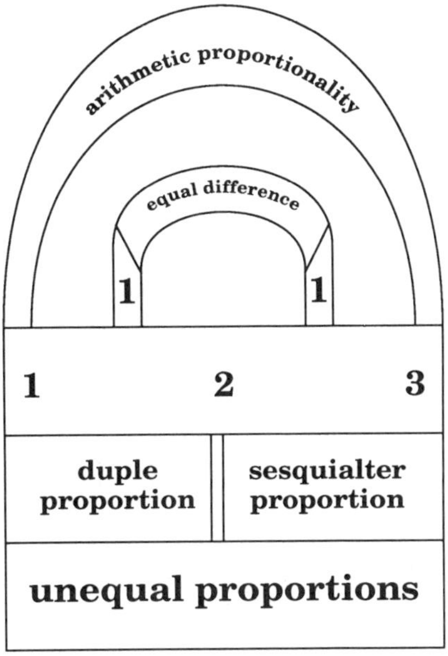

Figure 3.7.2

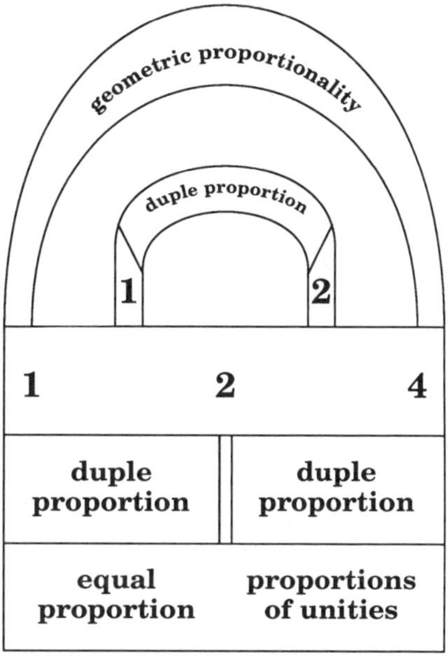

Figure 3.7.3

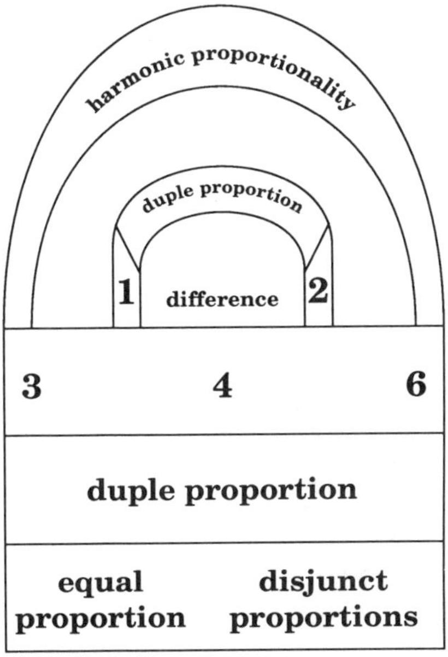

Boethius has drawn together in his *On Arithmetic,* since they may not be represented well enough in the proposed number. **36** Yet, since it has been shown more fully that inequality is derived from equality in the same manner **36a** in which all numbers are accustomed to be assembled from unity, **37** I consider that one hardly may pass in silence over the three proportionalities shown above, which from this very equality, in the same manner, produce nature by means of certain rules, [f1v] so that it leaves no inquirer behind in doubtful suspense. **38** Indeed, after the three equal terms have been arranged in two ways, the arithmetic mean is established. **39** Let the first [upper] term be equal to the first [lower], the second, to the first plus the second, the third, to the first plus the second plus the third, **39a** as here:

1 1 1
1 2 3.

40 Accordingly, the upper series is of unities. **41** For that reason, unity is placed under the first unity, which is equal to it; **42** 2 is placed below the second unity and it is equal to the first plus the second. **43** But placed below the third unity is 3, which is made up of these three unities, that is, the first plus the second plus the third in an equal distribution. **44** The same scheme will operate in the arrangement of three 2s or 3s with the substitution of equivalent terms in the same order. **45** Therefore, unity itself, constituted by the equality of the unities, will be the difference between the terms, for these terms omit no other terms between them. **46** But if the equality of the 2 is arranged, 2 will be the difference, **47** and one term is always omitted among these numbers. **48** But if the 3 will have distributed the equality, this [3] will make the difference, **49** and between the numbers two numbers naturally intervene, and so forth. **50** By "terms" in this discussion we understand numbers listed above and below by means of an equation.

51 A second way arithmetic proportionality is produced is by this progression: **52** after three equal numbers have been established [as the upper series] the following terms may be placed below them, the first of which is equal to the first plus the second [upper terms], the second to the first plus twice the second, the third to the first plus twice the second plus the third, **52a** as here:

36 Cf. above 3.5.
38–39 Boet. *De inst. mus.* 2.15.244.6–9.
39a Boet. *De inst. mus.* 2.15.244.12–15.
44 Boet. *De inst. mus.* 2.15.244.16–24 summarized.
45–49 Boet. *De inst. mus.* 2.15.244.25–32.
51–52 Boet. *De inst. mus.* 2.15.244.32, 245.1–4.
52 Cf. above 3.5.64–70.
52a–53, 55–56, 57, 57b–58 Boet. *De inst. mus.* 2.15.245.8–15.

1 1 1
2 3 4.

53 Here unity constitutes the difference between the terms, **54** for between 4 and 3 this unity makes the difference, **55** and between 3 and 2 it is the same. **56** And no natural number intervenes—**57** after 2, in fact, comes directly 3, **57a** and after 3, 4 **57b** follows in its natural disposition. **58** The same process will occur if I arrange three 2s; **59** then the difference of the terms will be 2, unity naturally intercepting the terms, **59a** as here:

2 2 2
4 6 8.

60 When, however, an arrangement of 3s is the basis of equality with the same imitation maintained, this will lead to a difference of 3 between the [lower] terms, **61** and two numbers will fall naturally between these terms, **61a** as is here distinctly perceivable:

3 3 3
6 9 12.

62 But if the numbers 4 or 5 are thus arranged, the same will happen, **63** and the number of intervening numbers will always be smaller by one than the difference between these [lower] terms; and so on for the remaining numbers.

64 Geometric proportionality proceeds from equality in this manner: **65** after three equal terms have been established [for the upper series], the first term placed below is equal to the first [upper term], the second is equal to the first plus the second, and the third, to the first plus twice the second plus the third, **66** as for example, if we list three unities [in the upper series] we get:

66a 1 1 1
1 2 4.

67 The same will happen if the arrangement is created out of the numbers 2 or 3 or from other numbers. **68** For in this series geometric proportionality

59 Boet. *De inst. mus.* 2.15.245.15–18 summarized.
59a Boet. *De inst. mus.* 2.15.245. 19–20.
60, 62 Boet. *De inst. mus.* 2.15.245.24–27.
64 Boet. *De inst. mus.* 2.15.246.1–2.
65 Boet. *De inst. mus.* 2.15.246.5–7.
66a Boet. omits this table in his *De inst. mus.* 2.15., perhaps because he already has included it in the previous discussion of the emergence of inequality from equality (cf. *De inst. mus.* 2.7). In the latter passage there is, of course, no reference to the geometric proportionality.
67 Boet. *De inst. mus.* 2.15.246.7–8 glossed.
68 Boet. *De inst. mus.* 2.15.246.8–9.

assumes beginning from equality. **69** The harmonic mean, as we shall elaborate more fully later, [f2r] arises from equality in the following manner: **70** after three equal numbers have been established [as the upper series], the first number placed below is equal to the first plus twice the second [upper terms]. **71** For example, if three unities have been arranged above, the first [lower] term is equal to the first plus twice the second unities; **72** the second is equal to twice the first plus twice the second unities, **73** and the third [lower term] is equal to the first [unity] plus twice the second plus three times the third, **73a** as here:

1 1 1
3 4 6.

74 And if the equality should reside in the numbers 2 and 3, the same reckoning of the mean will come out with regard to the terms and differences, which are separated by a duple. **75** If a triple proportion is sought equally between the extreme terms and between the differences, however, **76** let the first [lower] term be equal to the first plus the second [upper terms], the second to the first plus twice the second, and the third term to the first plus twice the second plus three times the third, as here:

1 1 1
2 3 6.

77 Also, the same scheme is applied to the numbers 2 and 3, and other numbers.

Chapter **Eight**
The proportions that lead to musical consonances

1 It is worthwhile to touch upon the nature of those proportions which bring forth consonances. **2** There are three in the multiple genus: the duple, the triple, and the quadruple. **3** There are also three in the superparticular genus: the sesquialter, the sesquitertial, and the sesquioctaval. **4** The duple proportion is nearest to equality, **5** and it is the first species of the multiple genus. **6** By nature this species is indivisible into two equal parts, **7** for it is

69 Cf. below 4.3.4ff.
69–73 Boet. *De inst. mus.* 2.15.246.13–18.
73a Boet. *De inst. mus.* 2.15.246.28–29.
74 Boet. *De inst. mus.* 2.15.246.24–26.
75–76 Boet. *De inst. mus.* 2.15.246.30–32, 247.1–4.
77 Boet. *De inst. mus.* 2.15.247.3–4 (second and third tables) summarized.
1–7 Boet. *De inst. arith.* 1.24.49.15–22 glossed.

made up of the first two superparticular proportions, that is, the sesquialter and the sesquitertial, which are unequal. **8** For these are the first proportions that divide the duple proportion, although the root of duple proportions is made easily. **9** The first proportion, the duple, arises from natural number, as the number 2 compared to unity yields it simply and by itself—**10** not from the mingling of the sesquialter and sesquitertial proportions, however. **11** Yet the sesquialter and sesquitertial proportions form the duple in the following order: 2, 3, 4. **12** [The number] 3 is compared to 2 in the sesquialter proportion, **13** and 4 and 3 are related in a sesquitertial proportion. **14** [The number] 4 confronts 2 in the duple proportion, **15** and sometimes, the duple proportion itself intercepts the other superparticular ratios beyond the sesquialter and sesquitertial, as is established in these numbers: **15a** 4, 5, 6, 7, 8. **16** Indeed, 6 related to 4 **16a** is shown to constitute a sesquialter proportion, 8 to 6 a sesquitertial. **17** Thus the sesquialter and the sesquitertial complete a duple proportion from 8 to 4. **18** Again, the natural progression of a duple proportion of this kind includes the other superparticular and superpartient ratios. **19** For 5:4 yields a sesquiquartal, 6:5 a sesquiquintal, 7:6 a sesquisextal, 8:7 a sesquiseptimal, **19a** 7:4 a supertripartientquartal [f2v], **19b** and [7]:5 a superbipartientquintal. **20** In like manner, 8:5 have been well known to round off a supertripartientquintal proportion very clearly. **21** Nevertheless, it is necessary to know that the first two superparticular ratios always divide this duple proportion as if they were the major and elementary parts. **22** For these major and elementary parts have been shown to divide and unite the first duple proportion **22a** in the natural numbers beyond the root [i.e., the denominator 1] of the duples—naturally, without interposing another number. **23** Truly, since a sesquialter proportion is larger than a sesquitertial, the former exceeds the true half of a duple proportion. **24** Therefore, a sesquitertial proportion does not fill up the entire half of this duple proportion. **25** Also, nature causes that whenever something is cut in such a way that it is not divided by equal parts, however much the smaller part is smaller than the half, by that much does the larger part, which is augmented, exceed the half. **26** Therefore, a sesquialter proportion is larger than a true half of a duple proportion by the same amount as a sesquitertial proportion comes out smaller than half of this duple proportion. **27** Therefore, if we arrange two successive sesquialter

10–14 Boet. *De inst. mus.* 2.10.240.25, 241.1–3.
15a Boet. *De inst. mus.* 2.10.241.5–6 altered.
16 Boet. *De inst. mus.* 2.10.241.6.
19 For a summary of this terminology, see above 3.3–4.
19a Boet. *De inst. mus.* 2.5.230.14–15; Boet. *De inst. arith.* 1.20.60.9.
19b, 21–22, 23–26 Boet. *De inst. mus.* 2.5.230.13–15 expanded.
27 Boet. *De inst. arith.* 2.2.81.7–10.

proportions, that which results from the extreme terms does not actually yield a duple proportion but rather exceeds it, as is established in this arrangement: 4, 6, 9. **28** Here 6:4 is a sesquialter proportion. **29** Likewise, 9:6 produces a sesquialter proportion. **30** Again, 9:4 does not accurately make a duple proportion but surpasses it. **31** Namely, the 8 has been included which is related to this 4 by the very duple proportion. **32** But if we compare 9 to 8, a sesquioctaval proportion is produced. **33** Two sesquialter ratios are also shown to exceed a duple proportion by one sesquioctaval proportion. **34** But since the sesquitertial proportion is smaller than the true half of a duple proportion, **35** for that reason two sesquitertial proportions can least complete the perfect duple proportion, as appears in these numbers: 9, 12, 16. **36** For 12:9 is a sesquitertial proportion, **37** and 16:12 brings forth a sesquitertial proportion. **38** But if we compare 16 to 9, they do not complete a duple proportion. **39** Surely, when 2 is added, which is the eighth part of 16, this duple proportion will be produced exactly, as is here established: 9, 12, 16, 18. **40** Observe that 18:16 produces a sesquioctaval proportion **41** and that 18:9 preserves a duple ratio; **42** so two sesquialter proportions exceed a duple proportion by one sesquioctaval proportion. **43** Two sesquitertial proportions [also] are smaller than a duple ratio by one sesquioctaval proportion. **44** Therefore, a sesquialter proportion surpasses a sesquitertial by the same sesquioctaval proportion, which is exhibited in these terms: 6, 8, 9. **45** For 8:6 makes a sesquitertial proportion, while 9:8 makes a sesquioctaval proportion. **46** Again 9:6 join in a sesquialter proportion. **47** Therefore, a sesquitertial proportion is smaller than a sesquialter by one sesquioctaval proportion.

48 A sesquioctaval proportion by nature is indivisible into two equal parts, **49** and to demonstrate this the terms must be arranged which bring together this same proportion, which derives from 9:8, because no number occurs naturally between them. **50** These numbers (the root of the sesquioctaval proportion having been omitted) are 16, 17, 18, **51** for 18:16 produces a sesquioctaval ratio. **52** But because the middle term to the smaller term, that is, 17:16, completes a sesquidecimasextal proportion, **53** and the larger number to the middle number, that is, 18:17, make up a sesquidecimaseptimal, **54** it follows that a sesquioctaval proportion itself can under no circumstance be separated or coupled in two equal parts. **55** For indeed, unequal ratios are formed from the extreme terms and the middle. **56** To be sure, a sesquidecimasextal proportion is larger than a sesquidecimaseptimal proportion. **57** But if a sesquialter is added to a duple proportion, the second species of this multiple genus will be produced, that is, the triple proportion, as here: 3, 6, 9. **58** For, indeed, 6:3 creates a duple proportion, 9:6 a sesquialter, and 9:3 a triple

48–56 Boet. *De inst. mus.* 1.16.202.17–27, 203.1–7.

proportion. **59** Therefore, because a sesquialter proportion surpasses a sesquitertial by one sesquioctaval, **60** if we connect a sesquitertial ratio to a duple, it cannot complete a triple proportion, as is demonstrated in these numbers: 3, [6], 8; **61** for 6:3 makes a duple proportion, **62** 8:6 a sesquitertial, and 8:3 can by no means reach a triple proportion. **63** But if a sesquioctaval proportion is added above, a triple proportion will directly respond to the 3, as occurs here: 3, 6, 8, 9. **64** For 9:8 places a sesquioctaval proportion on top, **65** and at the same time 9:3 comprise a triple proportion. **66** But if before the addition of the 9 which produces the sesquioctaval proportion a sesquialter proportion had arisen on top, **67** the third species of the multiple genus would have come forth, that is, the quadruple proportion, which these numbers show: 3, 6, 8, 12. **68** Observe that 12:8 corresponds to the sesquialter proportion, 12:6 to the duple, and 12:3 to the quadruple proportion. **69** But had I placed a sesquitertial ratio, which is produced from a duple and sesquialter, above the triple, **70** you would in like manner discern that a quadruple proportion is completed, as shown in these numbers: 3, 6, 9, 12. **71** For 12:9 shows a sesquitertial proportion, **72** and the same 12 holds with 3 a quadruple proportion. **73** All of this is perceived quite clearly in Figure 3.8.1 [f3v].

74 I consider it worthy to be known also that if a multiple proportion is multiplied by 2, then that which is produced will yield a multiple proportion. **75** But if that which proceeds from this multiplication is not a multiple proportion, then what was multiplied was not originally a multiple proportion. **76** In like manner, if a superparticular proportion is multiplied by 2, then what comes out of this multiplication will be neither a multiple nor a superparticular. **77** If that produced from such a multiplication is neither a multiple nor a superparticular, **78** then what was multiplied was either a superparticular or of another genus—however, not of the multiple. **79** It is evident from what has been said that the first two superparticular ratios, the sesquialter and the sesquitertial, bring about the first multiple, that is, the duple proportion. **80** Also it is evident that the first superparticular proportion together with the first multiple bring forth the second multiple, that is, the triple proportion. **81** The second multiple proportion, that is, the triple, and the second superparticular, that is, the sesquitertial, form the third multiple proportion, that is, the quadruple. **82** The other genera through one and another mixing of the multiple and superparticular proportions, which mutually correspond in the natural order, are naturally encapsuled in the subsequent multiple ratios. **83** If I juxtapose a sesquialter proportion and a duple I see a triple proportion result.

79 Cf. above 3.8.11–14.
80 Cf. above 3.8.57–58.
81 Cf. fig. 3.8.1.
83 Cf. above 3.8.57–58.

Figure 3.8.1

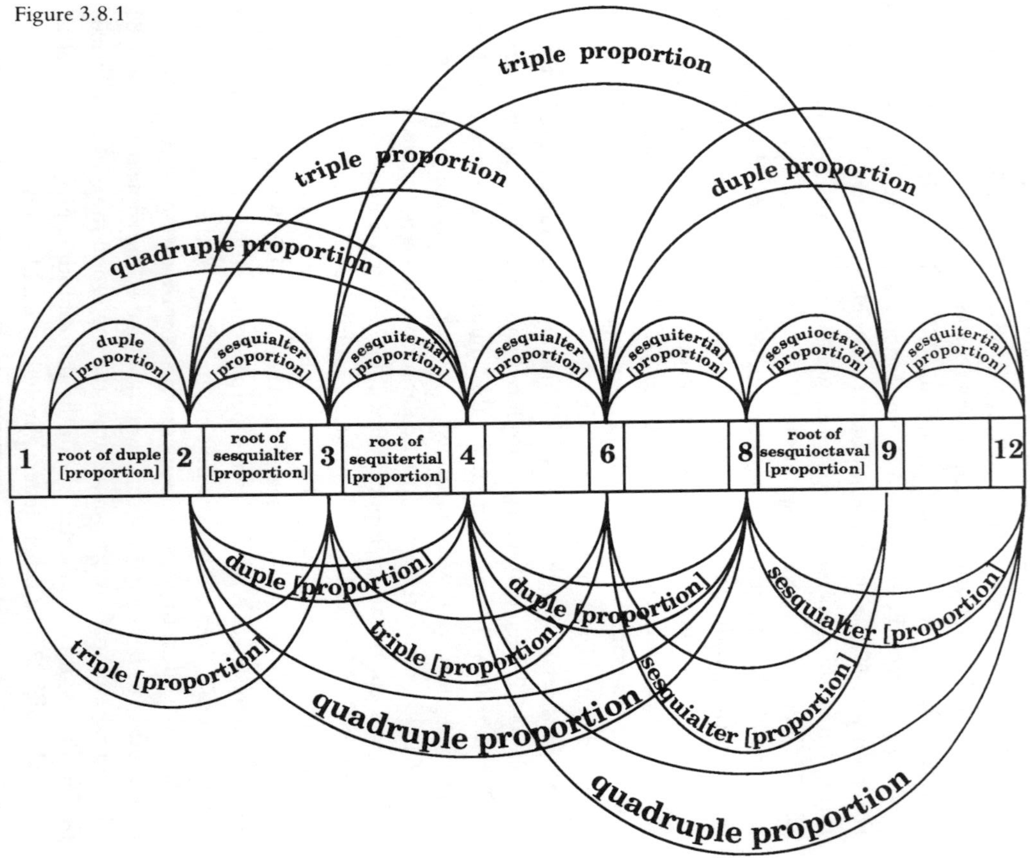
triple proportion
triple proportion
duple proportion
quadruple proportion
duple [proportion]
sesquialter [proportion]
sesquitertial [proportion]
sesquialter [proportion]
sesquitertial [proportion]
sesquioctaval [proportion]
sesquitertial [proportion]
1
root of duple [proportion]
2
root of sesquialter [proportion]
3
root of sequitertial [proportion]
4
6
8
root of sesquioctaval [proportion]
9
12
duple [proportion]
duple [proportion]
sesquialter [proportion]
triple [proportion]
triple [proportion]
sesquialter [proportion]
quadruple proportion
quadruple proportion

84 If I join a sesquitertial proportion to a triple, then a quadruple proportion will be formed. **85** If I join a sesquiquartal to a quadruple proportion, this will make a quintuplc, **86** and if I join a sesquinquintal proportion to a quintuple [f4r], this will undoubtedly achieve a sextuple proportion. **87** And the other multiple proportions are sought through the same process; **88** the same results in reverse, since naturally we bring together these multiple proportions with their coordinate superparticular proportions. **89** Nevertheless, notice that what is called the duple proportion in numbers becomes the consonance of the diapason in sounds, **90** and that called sesquialter or hemiolia is the consonance of the diapente in sounds, **91** whereas that which is the consonance of the diatessaron in sounds we call in numbers the sesquitertial or epitrite proportion. **92** The sesquioctaval proportion in numbers is called the whole tone in sounds. **93** Moreover, that which in numbers is the triple proportion is called the consonance of the diapason-plus-diapente in sounds, **94** whereas the quadruple proportion in numbers is measured and named bisdiapason in sounds. **95** The individual proportions will be dealt with in more detail in the following book.

84–86 Cf. fig. 3.8.1.

BOOK **FOUR**

Chapter **One**
The consonances produced from the proportions

1 Out of the great variety of numbers, Macrobius, Boethius, and other musicians are known to have used only six, relating them to numbers smaller than 6, from which the musical consonances are shown to be derived. **2** It is nevertheless necessary (even though previously mentioned) to show by a clearer demonstration that there are three multiple ratios and three superparticular. **3** The multiples are the duple, triple, and quadruple. **4** The superparticulars are the "hemiolic" or the sesquialter, the "epitrite" or the sesquitertial, and the "epogdous" or the sesquioctaval. **5** A duple number is one which, when compared with a smaller number, contains it twice; **6** the proportion, such as 4:2, is called duple. **7** A triple number is said to be that which, related to a smaller number, contains it three times. **8** The proportion, **8a** such as 6:2, **8b** is called triple. **9** A quadruple number is one which, compared with a smaller number, contains it four times; **10** the proportion, such as 8:2, is called quadruple. **11** A hemiolic number is one which, considered in relation to a smaller number, contains this whole number once plus a half part, such as 3:2. **12** They [i.e., the above-named writers] suppose that "hemiolic" is derived from "hemi," meaning half, and "olon," whole; that is, a whole and a half. **13** "Sesquialter" is appropriately derived from "sesqui," which is the half (as Cola Montanus had it) **13a** and from "alter," since a part equal to half the other number is added to the integer. **14** This proportion itself is the hemiolia or the sesquialter. **15** An epitrite number is one which, when compared with a smaller number, contains its entirety [f4v] and its third part, such as 4:3. **16** Certainly I consider "epitrite" to be derived from "epi," (above), and "tritos," (third,) that is, the larger number surpasses the smaller number by its third part. **17** Likewise, "sesquitertius" is derived from "sesqui," which is the half,

3–11 Macrob. *Somn. Scip.* 2.1.15–18; D'Abano *Expo. Pr. Arist.* 13.
5 Cf. above 3.3.16–16a.
7–8 Cf. above 3.3.21–21a.
9–10 Cf. above 3.3.24.
13 Cf. also above 3.3.33–34.
15 See above 3.3.55a–56.
15–20 D'Abano *Expo. Pr. Arist.* 13.
17 Cf. above 3.3.37ff.

as already explained, and from "tertia," [third], **18** since a part equal to a third of one number is added to the other number, such as 4:3; such a proportion is called epitrite or sesquitertial. **19** An epogdoic number is one which, when compared with a smaller number, exceeds and prevails over it by the smaller number plus an eighth part of it. **20** [The word "epogdous"] is derived from "epi," (above), and from "ogdo," (eight); and this number is called sesquioctaval, such as 9:8. **21** Hence, the proportion is also called epogdoic or sesquioctaval. **22** The words hemiolius, epitrite, and epogdous are also the names of the larger terms in the sesquialter, sesquitertial, and sesquioctaval proportions, but the former are not the names of the proportions themselves. **23** Truly, since the proportions themselves receive the designations from the larger terms, **24** the same names now and then are attributed to these proportions, which are assigned to their larger terms.

25 Moreover, attention should be called to the fact that musical consonances are produced from the proportions of greater inequality, **26** as the philosopher discloses, that is, from the multiples and superparticulars, **27** as has already been touched upon more broadly in the preceding explanations and in the Pythagorean investigation, **28** for the diapason results from the duple proportion, **29** the diapente from the sesquialter, and the diatessaron from the sesquitertial. **30** The tone results from the sesquioctaval proportion, the diapason-plus-diapente from the triple, and the bisdiapason from the quadruple. **31** The Pythagoreans and Platonists, therefore, conclude that the musical consonances are derived and produced from the multiple and superparticular proportions.

32 It appears that Aristotle himself readily agrees with these writers in Problem 41 of the section on harmony. **33** However, Ptolemy relies on the multiple-superpartient proportion to count the diapason-plus-diatessaron among the other consonances, as has already been stated above. **34** Thus it seems worthwhile to recommend these proportions to the faith of the reader; **35** by this let the reader think of the diapason as a consonance in the duple proportion, **35a** the diapente in the sesquialter, and the diatessaron in the sesquitertial, **35b** and the remaining consonances in the appropriate ratios. **36** These consonances need no defense, since they are axioms of music; **37** no, rather, they must be presupposed, as Aristotle established in the sixth [book]

25 Cf. also above 3.3–4, in particular 3.4.31.
27 Cf. above 3.4.31, 38, 89–91.
28 Cf. above 3.8.89.
28–32 D'Abano *Expo. Pr. Arist.* 41.
29–30 Cf. above 3.8.90–94.
33 Cf. above 2.5.40–59.
36–37 Barb., trans. *Them. Paraphr. (Metaph.)* 6.

of his *Metaphysics*. **38** In fact, among the Pythagoreans, it was the custom never to require any reason for what was laid down by their teacher Pythagoras, with solely the authority of the teacher implanted in them. **39** And it is permissible to consider those things which the minds of students, strengthened so long by a very firm doctrine, discover the reason for even without a teacher; **40** and furthermore, Saint Jerome found the teachable mind, even without a teacher, commendable and worthy of praise. **41** Nevertheless, there will be a clearer demonstration of the individual consonances in which each will be naturally derived from its appropriate proportion. **42** And in spite of mistaken and altered methods of inquiry will be traced the nature of consonances.

Chapter **Two**
The nature and the formation of the consonances from the proportions

1 Since every consonance is known to respect and conform with the nature of its own proportion from which it is produced, **1a** it is necessary that the consonance of the diapason preserve the property of the duple proportion. **2** Thus the diapente holds to the sequialter proportion, **2a** the diatessaron to the sesquitertial, **2b** the diapason-plus-diapente to the triple, **2c** the bisdiapason to the quadruple, **2d** and the tone to the sesquioctaval. **3** Therefore, what has been proposed concerning these proportions is likewise observed with greatest fidelity concerning their consonances. **4** In fact, continuous harmonic proportionality is to be displayed, so that the proposed consonances may be investigated and explained more easily and more clearly from these proportions.

5 In figure 4.2.1, [the ratio] 4:3 **5a** shows the sesquitertial proportion, **5b** presenting the consonance of the diatessaron, **6** and 6:4, **6a** the sesquialter, from which the consonance of the diapente is produced. **7** But the [ratio] 6:3 draws out the consonance of the diapason through the duple proportion, which the differences are known to bring together from the duple ratio. **8** Thus the diapente and the diatessaron divide and join in the diapason, which is comprised of two unequal parts, **9** just as the sesquialter and the sesquitertial

40 Hieron. *Dial.* 1.22.
1a Boet. *De inst. arith.* 2.48.156.22–23.
2 Boet. *De inst. arith.* 2.48.156.16–19.
2a Boet. *De inst. arith.* 2.48.156.4–9.
2b Boet. *De inst. arith.* 2.48.157.1–2.
2c Boet. *De inst. arith.* 2.48.158.4–6.
4–5, 5b–6, 7 Boet. *De inst. mus.* 2.16.247.8–14.

Figure 4.2.1

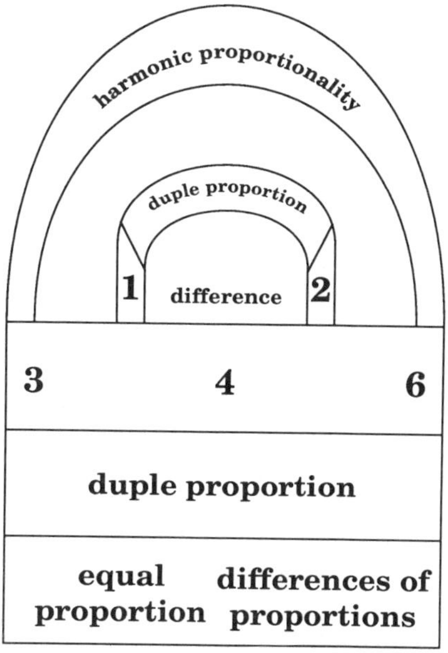

proportions are shown to bring together and divide the duple proportion. **10** But if we multiply the extreme terms one after another, and the middle [number] increases by its multiple, **11** the tones retain this mutual configuration in the sesquioctaval product, **11a** for 6 taken three times produces 18; **12** and similarly, 3 taken six times produces 18; **13** moreover, 4 multiplied by 4, that is, 4 squared, amounts to 16. **14** Therefore, it is established that 18 and 16, derived and gathered together one with the other, fill up the sesquioctaval proportion, which determines the tone. **15** Again, the smallest term, that is, 3, multiplied by itself produces 9. **16** The largest term, that is, 6, multiplied by itself produces 36, which, compared to 9, makes the quadruple proportion, which determines the bisdiapason. **17** When we consider these ratios more carefully, we will find in all a multiplication of terms or differences, for if the smallest term is multiplied by the middle term, 12 results. **18** Likewise, if we multiply this smallest term by the largest, it will make 18. **19** To be sure, the

10–11, 13 Boet. *De inst. mus.* 2.16.247.20–23.
14 Boet. *De inst. mus.* 2.16.247.23–24 paraphrased.
15 Boet. *De inst. mus.* 2.16.247.24–25.
16 Boet. *De inst. mus.* 2.16.247.26–28 paraphrased.
17 Gaf. here presumably meant "a multiplication of terms that are different."
17–22 Boet. *De inst. mus.* 2.16.247.28–29, 248.1–8.

Figure 4.2.2

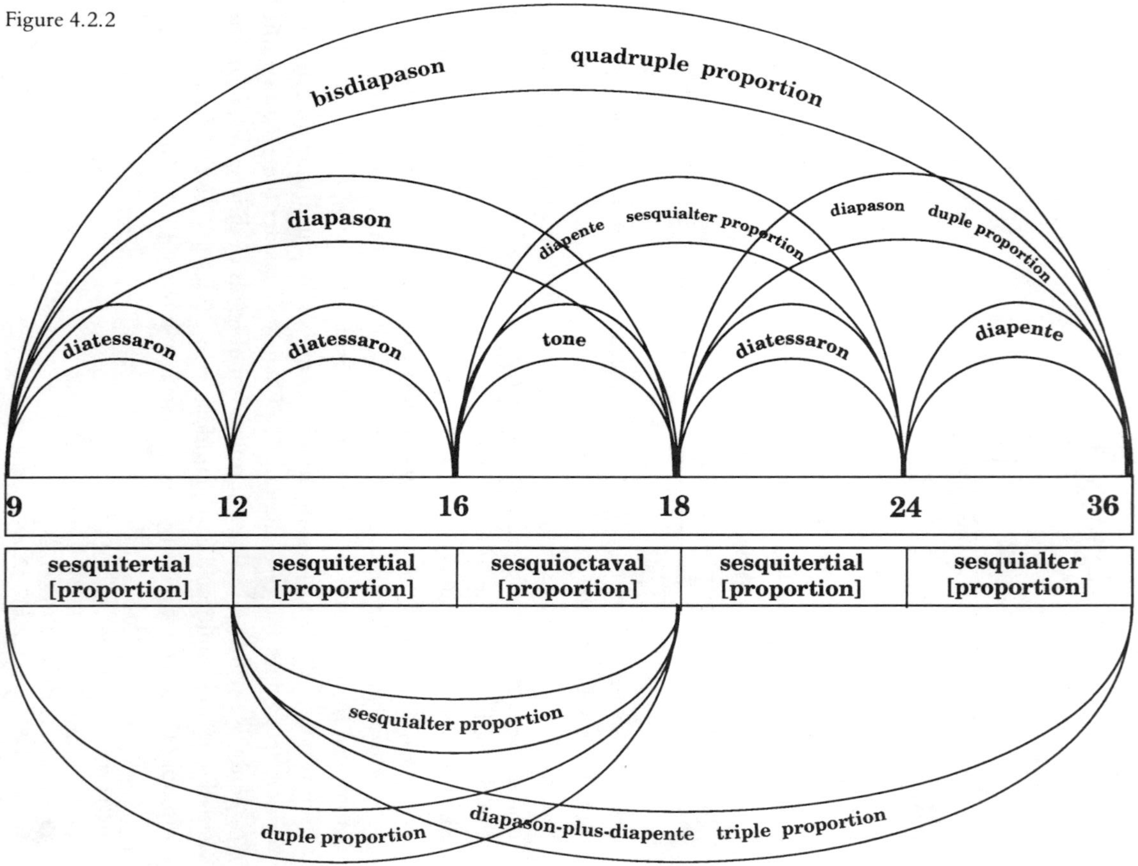

bisdiapason
quadruple proportion
diapason
diapente
sesquialter proportion
diapason
duple proportion
diatessaron
diatessaron
tone
diatessaron
diapente
9
12
16
18
24
36
sesquitertial [proportion]
sesquitertial [proportion]
sesquioctaval [proportion]
sesquitertial [proportion]
sesquialter [proportion]
sesquialter proportion
duple proportion
diapason-plus-diapente
triple proportion

middle term, multiplied by the numerical quantity of the largest term, amounts to 24. **20** Again, the smallest term multiplied by itself produces 9. **21** Likewise, the middle term multiplied by itself makes 16. **22** But 6, which is the largest term, arrives at 36 if it is multiplied by itself. **23** Therefore, 24:18, and in the same manner 12:9, yield the consonance of the diatessaron in the sesquitertial proportion.

24 But 18:12, and in like manner 24:16, **24a** yield the diapente **24b** in the sesquialter proportion [f5v]. **25** The diapason is produced from 36:18 and 18:9 in the duple proportion. **26** The diapason-plus-diapente is made from the triple proportion, that is, 36:12. **27** But the bisdiapason occurs in the quadruple proportion, which is made from 36:9. **28** Moreover, the tone arises from 18:16 **28a** by the sesquioctaval product. **29** Figure 4.2.2 shows all this very clearly.

30 A diligent investigator can easily perceive from figure 4.2.2 that the consonances one by one naturally correspond to their own proportion. **31** Nevertheless, it must be noted that if another consonance is added to the consonance of the diapason, which is acknowledged to reside in the duple proportion, **32** [the other consonance] will always observe the process and the order of the multiple, **33** so that, although this diapason exists in the duple proportion **33a** and the diapason-plus-diapente in the triple, **34** the diapason-plus-diatessaron can in no way be constituted in the order of multiplicity; **35** naturally [this is the case], for if the diapason-plus-diatessaron were placed in the multiple, the resultant proportion would have to be larger than the duple proportion and smaller than the triple, **36** whereas the diapason-plus-diatessaron is larger than the diapason and smaller than the diapason-plus-diapente.

37 But between the duple proportion and the triple no proportion intervenes naturally. **38** Therefore, the diapason-plus-diatessaron cannot be placed in the order of the multiple. **39** Thus the Pythagoreans [f6r] and also the Platonists feel that this is not a consonance.

Chapter **Three**
The nature of the tone and the semitones

1 Even if a consonance is subject to the nature of the proportion which produced it, it is necessary to determine the tone in the sesquioctaval propor-

23–24, 26–28 Boet. *De inst. mus.* 2.16.248.10–12.
33–37 Boet. *De inst. mus.* 2.27.259.27–30, 260.1–5.
38 Boet. *De inst. mus.* 2.27.259.15–16 paraphrased.

tion. **2** The result is that the tone is divided into two unequal semitones, since the sesquioctaval proportion itself is divisible into two unequal ratios; **3** for of the semitones one is smaller, the other larger. **4** For this reason, they are not fittingly called semitones, that is, things that contain an absolute half of the tone; **5** they are so called because they do not advance to a whole tone, just as the semivowel is not accustomed to be accepted in letters in place of middle vowels, **6** since indeed, it is proper that *semus* be preserved instead of *imperfectus,* for *semus, -a, -um* is the same thing as *imperfectus, -a, -um.* **7** The smaller semitone does not reach the half of the true tone. **8** In fact, the larger semitone exceeds the half of the tone itself by the very small particle by which the smaller semitone is surpassed by this half of the tone. **9** The Pythagoreans call the smaller semitone a diesis. **10** Plato, on the other hand, calls [the smaller semitone] a limma. **11** Later theorists, however, called it a semitone and labeled its half a diesis. **12** Nevertheless, some consider the half of the diesis as the comma. **13** Certain people, however, believe that the comma itself exceeds the half of the diesis, which Boethius teaches in the third book of his *On Music.*

14 Truly, Philolaus, the Pythagorean, transmits that the comma is divided in two "schismata" **14a** and also that the diesis is divided in two "diaschismata." **15** Whereas the smaller interval—which is harmonious **15a** according to Bacchius, a recent authority among the Greeks—**15b** is the diesis, that is,

2 Boet. *De inst. mus.* 1.16.202.17–21; 1.33.223.17–18; 3.1.269.31–32; 4.7.323.22–25 glossed.
2–3 Cf. above 3.8.48–56.
3 Boet. *De inst. mus.* 1.16.203.10–11; 1.33.223.20–21; Ugol. *Decl.* 5.13; 5.14.
4 Boet. *De inst. mus.* 2.29.263.5–7; 3.1.271.12–14; 3.1.272.7–9 glossed.
5 Boet. *De inst. mus.* 2.28.260.21–23 glossed.
7 Ugol. *Decl.* 5.13.2,4.
8 Ugol. *Decl.* 5.14.3.
9–10 Boet. *De inst. mus.* 2.28.260.21–25 glossed.
11 Boet. *De inst. mus.* 3.5.277.3–4 glossed.
12 Gaf. presumably is thinking of Greek theorists like Theon of Smyrna, who, in his work on mathematical sciences, writes in the second book ("On Music") that "the diesis according to the School of Aristoxenus is the quarter-tone" (cf. Michaelides, *The Music of Ancient Greece,* p. 81). Given that the first Latin edition of this work appeared only in 1644 (cf. Michaelides, *The Music of Ancient Greece,* p. 331), it is not quite clear how Gaf. would have had access to this information.
13 Boet. *De inst. mus.* 3.6.277.20–25 glossed.
14 Boet. *De inst. mus.* 3.8.278.15–17 glossed. The particular fragment of Philolaus, referred to by Boet., represents a *unica* (cf. Bower, trans., *Boethius: Fundamentals,* p. 96).
14a Boet. *De inst. mus.* 3.8.278.27–28 glossed. Gaf. substitutes "diesim" for Boet.'s "semitonium minus."
15, 15c Bac. *Intro.* 8. Gaf. refers to the same passage above in 2.3.8–9.
15b, 16 Gaf. here refers to Aristox., who in *Harm.* 1 states that "the voice cannot differentiate, nor can the ear discriminate, any interval smaller than the smallest diesis" (cf. Mi-

the half of the smaller semitone, **15c** which the later users of the diatonic genus reject, **16** they naturally regard the interval of the smaller semitone as the smallest interval. **17** We shall omit these smaller portions themselves of the tone or semitone, which have been examined on a broader basis in their own proportions by our Boethius and by Ugolino, his imitator. **18** In bringing together this musical property in the diatonic genus alone, the smaller semitone is accordingly composed of two dieses, whose intervals result naturally in the arrangement of the harmony, either before or after or in between two tones, that is, as it is said in the investigation of the diatonic genus.

19 The apotome, on the other hand, which means "cutting off," is called the larger semitone by all, **20** and it is made from two dieses and a comma. **21** Indeed, the apotome exceeds the smaller semitone by the comma itself. **22** In fact, some erroneously consider that what occurs either before or after or in between two tones is the larger semitone, as is understood by many explanations. **23** For if we take away two tones from the consonance of the diatessaron, the smaller semitone remains; **24** accordingly, whenever we remove two sesquioctaval [proportions] from the sesquitertial proportion, that which remains will be the proportion of the smaller semitone.

25 After two adjacent sesquioctaval proportions have been placed, the origin and relative position of which must necessarily be sought in the second octuple proportion, **25a** let us arrange these numbers in this way: 64, 72, 81. **26** Certainly, two sesquioctaval proportions are contained within these numbers [f6v], **27** and, consequently, two tones are produced in the first natural arrangement. **28** Now, that number which is related to the smaller number of those arranged in the sesquitertial proportion, that is, 64, must be added. **29** But because 64 lacks a third part, if we multiply all these numbers, so arranged, by 3, even this smaller number will appear, resolvable into three equal parts. **30** And all the numbers will maintain the same proportion they had before they were multiplied by 3. **31** So 64 taken three times results in 192,

---

chaelides, *The Music of Ancient Greece,* p. 81). Aristid. Quint. (*De Mus.* 1.7) holds the same viewpoint (cf. Mathiesen, trans., *Aristides Quintilianus: On Music,* p. 80).

17 Cf. Boet. *De inst. mus.* 3.11–16; 4.5–12; Ugol. *Decl.* 5.13.19–28, 40, 48, 52, 68–69; 5.15.5–18, 24–32.

19 Boet. *De inst. mus.* 2.30.263.21–22; Ugol. *Decl.* 5.14.2.

20 Boet. *De inst. mus.* 3.8.278.17–19.

23 Boet. *De inst. mus.* 1.33.223.19–20 glossed. In his description of the components of the diatessaron, Boet. mentions "duobus tonis semitonioque consistere." Gaf. is more specific in his classification of the semitone as a "semitonium minus."

24 Boet. *De inst. mus.* 2.28.260.25–28 glossed. As in 23 Gaf. again is more specific in his statement than Boet.

25 Boet. *De inst. mus.* 2.28.261.5–7.

25a, 28–30 Boet. *De inst. mus.* 2.28.261.10–17.

31, 34–35 Boet. *De inst. mus.* 2.28.261.17–25; Ugol. *Decl.* 5.17.12.

whose third part, that is, 64, when added to it, is shown to complete the number 256. **32** Then 256:192 will be the sesquitertial proportion, which determines the consonance of the diatessaron. **33** Now let us arrange two adjacent sesquioctaval proportions, preserving the smaller number, that is, 192, in its proper order. **34** Then 72 taken three times produces the number 216. **35** On the other hand, 81 taken three times results in the number 243, which is arranged in the following manner between the two previously mentioned numbers that comprise the sesquitertial proportion. (cf. fig. 4.3.1).

36 After these numbers have thus been arranged, the number farthest from the first produces the sesquitertial proportion, observing the consonance of the diatessaron. **37** However, the second number to the first produces the sesquioctaval proportion and thus the tone. **38** In like manner, the third number is related to the second in the sesquioctaval proportion, comprising the tone. **39** The space which extends from the number 256 [g1r] to the number 243 represents the first form of the smaller semitone, which has been investigated and derived from these first smallest numbers. **40** But as it has been laid down as a law beforehand, that this smaller semitone does not fill the proper half of the tone, I have considered that this be proven by a more evident explanation. **41** Because the difference between the numbers 243 and 256 amounts to only 13 units, **42** if we multiply this eighteen times, it will not yield this whole smaller number but will fall short of it, for 234 will result, which is smaller than 243. **43** When the difference, that is, 13, is multiplied nineteen times, **43a** the product will exceed this smaller number, that is, 243, for it will yield 247. **44** So this difference comprises less than the eighteenth part of the smaller number itself, but, in fact, more than the nineteenth part.

45 It is necessary that every semitone be placed between 17:16 and 18:17, if it is to preserve the integral half of the tone, which is, in fact, impossible, **46** for, as has been shown above, **46a** the sesquioctaval proportion is divided into two unequal parts, **47** as that which occurs between 18 and 16, since the number 17 lies naturally in between. **48** This is so because 17:16 does not preserve the same proportion which 18:17 shows. **49** Thus, the sesquioctaval proportion itself cannot be divided into two equal semitones. **50** But if it were

36–37 Boet. *De inst. mus.* 2.28.261.25–28; 1.17.203.20–25.
38 Boet. *De inst. mus.* 2.28.261.27–28.
39 Boet. *De inst. mus.* 2.28.261.28–30. Gaf. has reversed Boet.'s order of the numbers 256 and 243.
41–43 Boet. *De inst. mus.* 1.17.204.1–8. The error in lines 3 and 6 ("octies" instead of "deciesocties") is corrected in the later passage; see Boet. *De inst. mus.* 2.29.262.5–10.
46 Cf. above 3.8.48–56, 4.3.2–3.
46a–47 Ugol. *Decl.* 5.12.11.
48 Ugol. *Decl.* 5.12.16.
49–51 Ugol. *Decl.* 5.12.18–19 paraphrased.

Figure 4.3.1

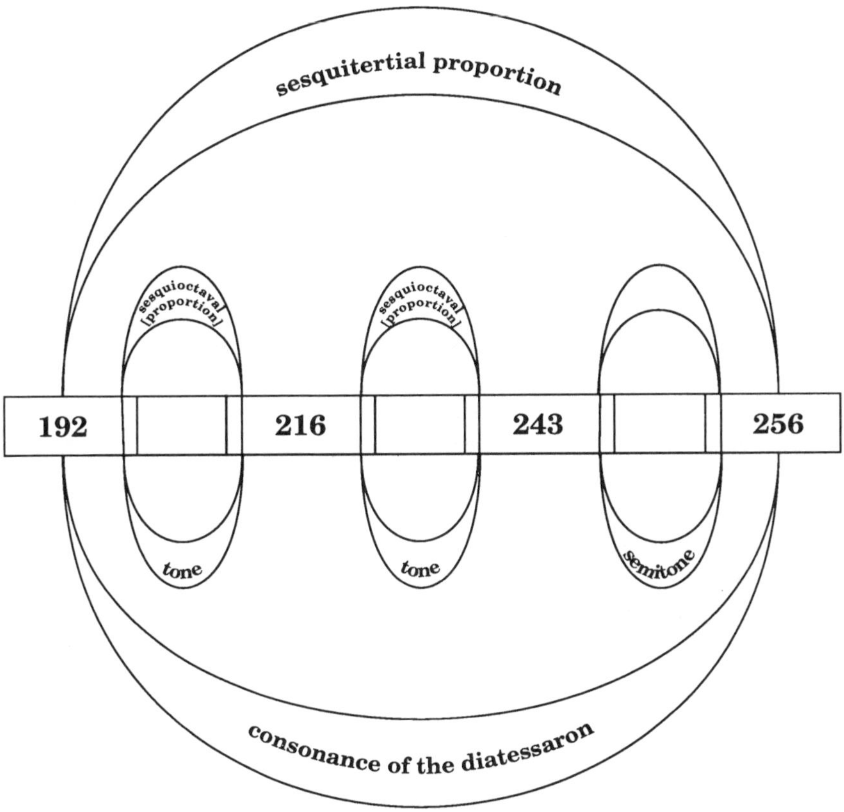

attempted, it would be necessary to take away a small portion from the 17:16 proportion, which must be added to the 18:17, from which they would behave equally, **51** what cannot be done under any circumstances, since the 17:16 does not contain any smaller part in itself than the 17:16, that is, the indivisible union, **52** which one strives to preserve completely and inviolably. **53** Therefore, it is agreed in a clearer way that the tone cannot be divided into two equal semitones. **54** The smaller semitone, however, does not reach the middle of the tone; **55** therefore, if the semitone is doubled, this double cannot complete the tone. **56** Thus two smaller semitones cannot complete the full interval of the tone. **57** In fact, it naturally happens that when any sort of half is duplicated, it produces the complete whole of which before duplication it was

52 Ugol. *Decl.* 5.12.18–19 paraphrased.
53–56 Boet. *De inst. mus.* 2.29.263.2–5 paraphrased.
57 Boet. *De inst. mus.* 2.29.263.5–6 paraphrased.

a half. **58** If, however, the duplication does not complete the whole, then it follows that what is duplicated was not the half. **59** To be sure, if the duplicated particle rises beyond its whole, then before it was duplicated it was larger than the half part. **60** Therefore, that part which is truly called a semitone among the consonances is smaller than the exact half of the tone; **61** however, that which remains of the whole tone is larger than half of this tone. **62** Therefore, this part is called the larger semitone.

63 By as much as the smaller semitone is smaller than the exact half of the tone, by so much the apotome exceeds the unimpaired half of the tone. **64** And since it has been shown that the semitone lies principally between 256 and 243, **64a** it must now be discussed in what larger numbers the apotome exists. **65** In fact, if 243 can accept an eighth part so that its sesquioctaval proportion can be compared with it, namely, 256, **66** this representation, having been placed in relation to the sesquioctaval proportion of this smaller number, would define the apotome by a necessary ratio. **67** But since this number 234 cannot have an eighth part [g1v], let the individual numbers rise by the [factor] eight; **68** and certainly, the number 243, taken eight times, completes the number 1944, **69** to which if its own eighth [part], that is, 243, is now added, 2187 results. **70** Likewise, let the number 256, taken eight times, grow to 2048; **71** [this number] will bring together that number **71a** which is placed as the middle number of those previously discussed, as in figure 4.3.2.

58 Boet. *De inst. mus.* 2.29.263.6–7 paraphrased.
59 Boet. *De inst. mus.* 2.29.263.7–9 paraphrased.
60 Boet. *De inst. mus.* 2.29.263.18–19 paraphrased.
63–64 Boet. *De inst. mus.* 2.30.263.26–28, 264.1–2: "Quantum igitur semitonium minus integro dimidio toni minus est, tantum apotome toni integrum superat dimidium. Et quoniam docuimus semitonium in CCLVI et CCXLIII principaliter stare, nunc ea, quae apotome dicitur." Gaf. here obviously confused the terminology in both versions of the treatise: ("in quibus minoribus numeris constet apothome" instead of the correct "nunc in quibus maioribus numeris constet apothome").
64 Gaf.'s discussion here as well as in the preceding sentences centers around the *semitonium minus* (also known as *limma* or *diesis*). The numbers 256 and 243 represent the ratio of the diesis; this proportion is arrived at through subtraction of two whole tones (ratio 9:8) from the interval of the diatessaron (ratio 4:3). Recalling that the subtraction of an interval is achieved through multiplication with its inverted fractions, we obtain:

$$\frac{4}{3} : \left(\frac{9}{8}\right)^2 = \frac{4}{3} \times \left(\frac{8}{9}\right)^2 = \frac{4}{3} \times \frac{64}{81} = \frac{256}{243}.$$

64a Gaf. here has clearly confused the Pythagorean terminology. In both the 1480 and 1492 editions Gaf. equates the *apotome* with the *semitonium minus* (as becomes evident from his writing: "minoribus numeris" [1492]; "minimus numeris" [1480]). Yet the proportion which he provides in *Theorica* 4.3.64 is that of the *diesis* (or *limma*) to the *semitonium minus,* but *not* of the *apotome* (see explanation to *Theorica* 4.3.73). The subsequent figure in the *Theoricum* identifies the *semitonium minus* and the *apotome* correctly.
65–71a Boet. *De inst. mus.* 2.30.264.3–14.
71 That is, 1944, 2048, 2187.

Figure 4.3.2

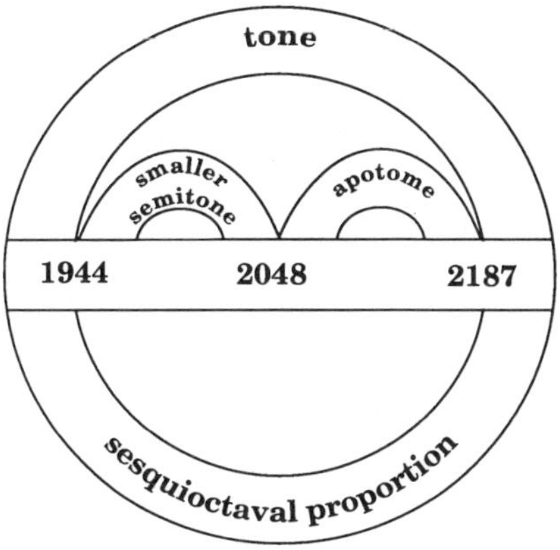

72 The third term to the first produces the tone, **72a** comprehended by the sesquioctaval ratio. **73** The second term to the first produces the smaller semitone, and the third term to the second **73a** the apotome **74** for the reason that since the larger proportion occurs in the smaller numbers, the larger consonance consequently results; **75** the smaller proportion, on the other hand, occurs in the larger numbers.

Figure 4.3.2 [as erroneously given by Gaffurio]

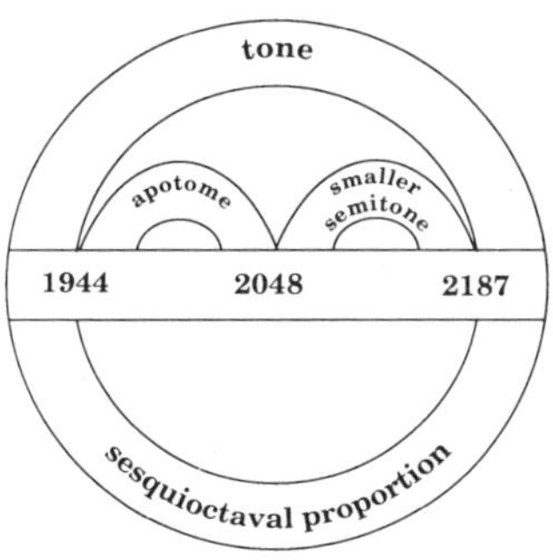

72 That is,

$$\frac{2187}{1944} = \frac{9}{8}.$$

Boet. *De inst. mus.* 2.30.264.15–16.

72a This is Gaf.'s insertion for clarification of 72.

73 The *apotome* is defined as the difference between the whole tone (ratio 9:8) and the *semitonium minus* or *diesis* (ratio 256:243). Applying a mathematical operation similar to the one in *Theorica* 4.3.64, we obtain:

$$\frac{9}{8} : \frac{256}{243} = \frac{9}{8} \times \frac{243}{256} = \frac{2187}{2048}.$$

76 In these primary numbers resides the proportion of the apotome, since the interval of the smaller semitone is comprised of 256 and 243, in the smallest [possible] numbers. **77** For that reason 1944 and 2048 are subject to the same proportion that binds 243 and 256, since the 243 and 256 are multiplied by 8. **78** To be sure, whenever one number is multiplied by any two numbers, then the numbers derived from the multiplication will be in the same proportion as had been those numbers which the multiplicator multiplied.

79 Nevertheless, Philolaus, the Pythagorean, shows the division of the tone more clearly, placing the first odd number, that is, 3, with its square, which is 9. **80** He multiplies this number by 3 and produces 27, which, when related to 24, **80a** preserves the tone in the sesquioctaval relation, **80b** whose difference certainly is 3; **81** 3 taken eight times, completes the number 24; **82** 3 taken nine times, surely completes 27. **83** He also divides the tone into a smaller semitone and an apotome, whose difference he called the comma, **84** declaring that the smaller semitone is composed of 13 units on that account, because between 256 and 243, the difference is foreseen, **85** and because likewise, this number 13 consists of the 9, 3, and 1, the last of which maintains an extraordinary place among the other numbers; **86** 3 and 1 are odd among the natural numbers, **87** and 9 is the first odd square number. **88** For that reason after having brought together the smaller semitone in the number 13, he proposed that the remaining part of 27, which is greater by 14 units, includes the apotome. **89** But because the [number] 1 evidently remains the difference between 13 and 14, it has been declared that this unit must be kept in the place of the comma, **90** since 1 produces the tone twenty-seven times in the units—**91** on that account because [these units lie] in between the numbers 243 and 216, which among each other are apart by the tone; **92** the number 27 itself will accommodate the difference.

73–73a Boet. *De inst. mus.* 2.30.264.16–17: "secundus vero ad primum semitonii minoris, apotomes vero tertius ad secundum." Gaf. confused the terminology in the *Theorica* ("Secundus vero ad primum apothomen sed tertius ad secundum semitonium minus" instead of the correct "Secundus vero ad primum semitonium minus sed tertius ad secundum apothomen"), whereas the *Theoricum* shows the correct identification of *apotome* and *diesis*.
73a In Pythagorean terminology, the *diesis (limma)* (ratio 243:256).
76–78 Boet. *De inst. mus.* 2.30.264.17–26.
79 Boet. *De inst. mus.* 3.5.276.15–17. Gaf. substitutes his "novenarius" for Boet.'s "primus cybum a primo inpari." In fact, Boet. resorts to "novenarius" in the subsequent passage (cf. *De inst. mus.* 3.5.277.8).
80, 80b Boet. *De inst. mus.* 3.5.276.18–22.
81–82 Boet. *De inst. mus.* 3.5.276.22–23, 277.1 paraphrased.
83–92 Boet. *De inst. mus.* 3.5.277.1–18.
85–87 For a fuller discussion of Pythagorean number theory, see *Theorica* 2.8.

Figure 4.3.3

A	diesis	B	diesis	C	diesis	D	diesis	E	comma	F
	Smaller Semitone			•		**Larger Semitone**				
	T			O		N		E		

93 From this, we may infer that the tone is composed of two semitones and one comma, just as the tone includes the apotome and the smaller semitone. **94** The semitone, in fact, differs from the apotome by a comma. **95** It is established that the apotome [consists of] nothing other than the smaller semitone [g2r] and the comma. **96** Hence, if we take away two smaller semitones from the tone, the comma is left, because this likewise results through the proof. **97** Since the diapason is contained in five tones and two smaller semitones, six tones exceed the diapason by one comma. **98** There is no doubt that after a group of five tones has been removed from the diapason, two smaller semitones remain. **99** Indeed, from the six tones, after having withdrawn a group of five tones, one tone remains, which surpasses these two semitones by a comma. **100** But when a comma is added to these two semitones, the tone will be completed. **101** Therefore, it is established that the tone is completed by two smaller semitones and a comma. **102** For indeed, the smaller semitone is brought together by two dieses. **103** On the other hand, the apotome or larger semitone comprises two dieses and the comma. **104** Then, the tone comprises four dieses and a comma, **104a** what the intervals of the letters [in figure 4.3.3] indicate.

Chapter **Four**
The nature of the diapente and the diapason

1 The consonance of the diapente, which resides in the sesquialter proportion, is composed of the diatessaron and of the tone, **2** for the sesquitertial and sesquioctaval proportions form the sesquialter. **3** Since the consonance of the diatessaron consists of two tones and a smaller semitone, a tone added to [the diatessaron] will complete the consonance of the diapente. **4** For this reason, let us arrange the numbers of the two sesquioctaval proportions **4a** and of the

93–96 Boet. *De inst. mus.* 3.6.277.20–25.
97–101 Boet. *De inst. mus.* 3.7.277.27–28, 278.1–8.
102 Cf. above 4.3.18.
103 Cf. above 4.3.20.
104 Boet. *De inst. mus.* 3.6.277.19 glossed.
1–4b Boet. *De inst. mus.* 2.31.265.1–3 expanded.

space of the smaller semitone **4b** in the formation of the consonance of the diatessaron **4c** as in figure 4.4.1: 192, 216, 243, 256. **5** The second term to the first and also the third to the second are in the sesquioctaval proportion; **6** hence, they show two tones. **7** The fourth term to the third occupies the space of the smaller semitone; **8** the same fourth term to the first completes the sesquitertial proportion, thereby producing the consonance of the diatessaron. **9** Now if I add to 256 its eighth part, 288 results, which answers to the number 256 as the sesquioctaval proportion and, consequently, the tone. **10** And the same 288, compared with 192, represents the sesquialter proportion, **10a** comprising the consonance of the diapente. **11** Thus, the diapente is composed of the diatessaron and the tone; **12** it is composed of three tones and a semitone, **12a** as figure 4.4.1 shows. [g2v]

13 In the arrangement of these numbers, the most remote term, that is, 288, [taken in relation] to the fourth term, that is, 256, renders the sesquioctaval proportion, as is clearly indicated, thence a tone. **14** Also, this most remote term to the first term, that is, 192, renders the sesquialter proportion; **15** for that reason, this proportion assembles the diapente. **16** Accordingly, it is clear that the diapente consists of three tones and a semitone; or, that it is composed from the diatessaron and the tone **17** because it has been confirmed that the sesquialter proportion is produced and generated from the sesquitertial and sesquioctaval.

18 Now it is necessary that the consonance of the diapason arises when the diatessaron and diapente joined together produce the ratio that the sesquitertial and the sesquialter proportions are accustomed to form, namely the duple. **19** For this reason it is clear that since the diatessaron is made up of two tones and a smaller semitone, and the diapente of three tones and a smaller semitone, **20** it is natural to conclude that the consonance of the diapason is made up of five tones and two smaller semitones. **21** But since these two semitones (as shown in the preceding chapter) do not complete the tone but exceed half of this tone, **22** the consonance of the diapason can under no circumstances comprise six tones **23** for, in fact, it falls short of six tones by the distance and interval of a comma. [g3r] **24** Truly, the comma, placed above these two smaller semitones, will complete the tone, **25** and in this way, the diapason would be made of six tones. **26** Consequently, the consonance of the diapason does not comprise six tones, **27** as Aristoxenus **27a** and Macrobius supposed,

4c–7, 9–10 Boet. *De inst. mus.* 2.31.265.5–11.
19 Boet. *De inst. mus.* 2.31.265.14–16.
21 Cf. above, fig. 4.3.1.
23 Boet. *De inst. mus.* 2.31.265.19–20.
26–27 Aristox. *Harm.* 2.56–58; Macrob. *Somn. Scip.* 2.1.25.

Figure 4.4.1

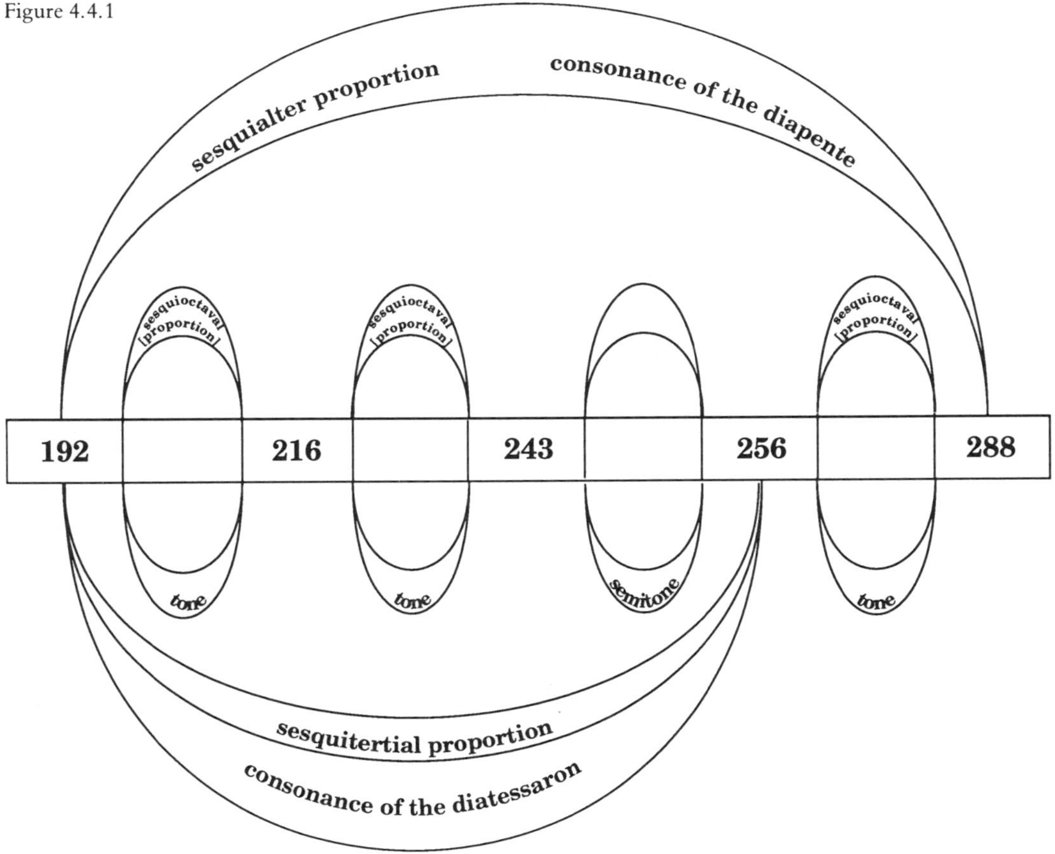
sesquialter proportion
consonance of the diapente
sesquioctaval [proportion]
sesquioctaval [proportion]
sesquioctaval [proportion]
192
216
243
256
288
tone
tone
semitone
tone
sesquitertial proportion
consonance of the diatessaron

in the second [book] of *The Dream of Scipio,* **28** and as even Pietro d'Abano, **28a** known as "the Conciliator" to physicians, does in [his commentary on] the thirteenth musical problem of Aristotle, **29** where he concludes that the bisdiapason possesses twelve tones.

30 Under these circumstances, it is proper that the following also be known. **31** If the consonance of the diapente is added to the consonance of the diapason, **32** another consonance will arise, named with the nouns of the two consonances, that is, diapason-plus-diapente, constituted in the triple proportion, as clarified above. **33** If then a diatessaron is added to this consonance of the diapason-plus-diapente, the consonance of the bisdiapason, which rests in the quadruple proportion, will directly result, as explained more clearly in the second chapter. **34** The diapason-plus-diapente consists of eight tones and three smaller semitones. **35** The bisdiapason is completed by ten tones and four smaller semitones, **36** as is well known to the more diligent inquirer and will be apparent afterwards in the division of the monochord. **37** However, if the sesquialter proportion is joined to the quadruple, the sextuple will then emerge, as is established in these numbers: 3, 12, 18. **38** Likewise, whenever the consonance of the diapente is added to the consonance of the bisdiapason, there will result another consonance named from both words [bisdiapason-plus-diapente]—**39** which musicians exclude from the classification of consonances because of the vast distance of the extreme sounds which alternately correspond to one another. **40** And since the multiple proportion—that is, the quintuple proportion—occurs naturally between the quadruple and sextuple, **41** let us attentively consider whether the sesquitertial proportion, which is smaller than the sesquialter, can take the place of a multiple, if it is added to the quadruple proportion.

42 Consequently, let us arrange numbers which assemble the quadruple and the sesquitertial proportions in the following order: 3, 12, 16. **43** [The relation] 12:3 completes the quadruple proportion. **44** On the other hand, 16:12 completes the sesquitertial proportion. **45** On the contrary, 16:3 does not yield the quintuple proportion exactly but is judged to exceed it by 1, which is the third part of 3. **46** But because no multiple proportion can naturally intercede between the quintuple itself and the sextuple—produced previously from the quadruple and the sesquialter—**47** the sesquitertial therefore cannot be added to the quadruple proportion, as the latter naturally resides in the multiple genus; **48** and in keeping very much to this, the

28–29 D'Abano *Expo. Pr. Arist.* 13.
33–34 Cf. above 4.2.2c, 27; also fig. 4.2.2.
35 Cf. fig. 5.3.1.
39 Cf. above 4.8.20.
48 For example, Boet. *De inst. mus.* 5.10–12 (cf. *Theorica* 2.5.23ff.).

musicians deny that the bisdiapason-plus-diatessaron can be a consonance, **49** and this, although it is derived from the proportion 16:3, which is the quintuple-sesquitertial proportion in the multiple-superparticular genus. **50** This bisdiapason-plus-diatessaron nevertheless cannot remain in the multiple genus, which is simple, **51** since in all quadruple proportions the order of multiplicity is retained, **52** and so is the order of superparticularity, since the sesquitertial proportion itself has been added to the quadruple, as is evident in these numbers: 6, 24, 32. **53** The quadruple proportion is made from 24:6, the sesquitertial proportion, from 32:24. **54** Also the other [proportions] are brought together, using 12, 48, and 64 coupled in pairs to one another. **55** Because the superparticular genera, joined to the multiple genera [g3v], therefore in a natural way bring together other multiple genera, in these this order and process will be overlooked, **56** as likewise in the duple and sesquitertial proportions, **56a** from which the diapason-plus-diatessaron is derived, as has been previously shown. **57** Thus, the bisdiapason-plus-diatessaron must not be reckoned among the other consonances. **58** Indeed, the diapason-plus-diatessaron and the bisdiapason-plus-diatessaron differ in this, **59** because the former is found in the multiple-superparticular genus, while the latter is present in the multiple-superpartient genus. **60** Since the extremes of the bisdiapason form with each other the quadruple proportion, the extent of the interval places this among the larger consonances. **61** To be sure, [the Greeks] call the diatessaron the smaller of the consonances, since the higher sound in this surmounts the lower by its third part. **62** Therefore, the kind of consonance that exists here can neither be extended beyond the quadruple proportion, nor compressed within the third part.

Chapter **Five**
The order of the consonances

1 In pursuit of the order of the consonances, the Pythagoreans decided that the consonance whose quality is more pleasant to the hearing and also whose nature turns out to be more open to both the senses and the mind, must be determined to be the first in the series. **2** Whatever exists of itself according to this musical perception, according to Boethius, is to be comprehended by the sense. **3** In this respect, since the consonance of the diapason, which stands in the duple proportion, is more familiar than the other consonances, we must conclude that it is the first of all consonances. **4** It deserves to be in a position superior to the remaining, which the diapason surpasses by virtue of a more

56 Cf. above 4.4.48.

lucid recognition. **5** The Pythagoreans themselves consider that the rest of the consonances that preserve the order of multiplicity—as the multiple proportions do not unjustly transcend the superparticular proportions—follow in the natural order. **6** The diapason is indeed the first [member] in the duple proportion; **7** the diapason-plus-diapente constitutes the second [member] in the triple proportion; **8** and the bisdiapason constitutes the third [member] in the quadruple proportion; **9** hence, let us arrange the natural numbers from the unity up to the 4 in this order: 1, 2, 3, 4. **10** [The number] 2 compared to 1 yields the duple proportion, producing the consonance of the diapason, which is the first and which is well known by its simplicity. **11** [The number] 3 compared to 1 shows the triple proportion, which brings together the diapason-plus-diapente. **12** On the other hand, 4, which is compared to 1 itself, yields the quadruple proportion, which preserves the consonance of the bisdiapason. **13** Because these multiple proportions perceive the 1 [unity], which is the first, and those multiples follow this 1 [unity] in their right and natural order, the Pythagoreans determined that these multiples take precedence and the superparticular proportions follow.

14 Now it remains to bring together the superparticular proportions themselves in the numbers which have been previously arranged above. **15** The ratio 3:2 preserves the sesquialter proportion, which the diapente maintains as a consonance—this, in fact, is the first consonance of the superparticular proportion in the natural arrangement of the numbers. **16** The ratio 4:3 yields the sesquitertial proportion, that is, the second superparticular proportion [g4r], which shows a consonance of the diatessaron. **17** The ratio 4:2 yields the duple proportion, just as 2:1. **18** Because the first two superparticular proportions bring about and divide this duple proportion, after the multiple proportions the Pythagoreans place these superparticular proportions, which yield the diapente and the diatessaron.

19 On the other hand, Eubulides and Hippasus decided on a different order of the consonances, saying that augmentation of multiplicity corresponds to diminution of superparticularity **20** in such a way that it is not possible for there to exist a duple without a half, **20a** nor a triple without a third. **21** Therefore, they placed first the duple proportion, that is, the diapason; **22** second, the sesquialter proportion, which yields the diapente as if it were in

14 Cf. above 3.5-6.

19-25 Boet. *De inst. mus.* 2.19.250.27–31, 251:1-9. Arist. *Top.* 135b 19f; Arist. *Metaph.* 1020b 26f (cf. Frieder Zaminer, "Konsonanzordnung und Saitenteilung bei Hippasos von Metapont: Wiederentdeckung eines frühen Lehrstücks," *Jahrbuch des Staatlichen Instituts für Musikforschung Preussischer Kulturbesitz* 1981–82, p. 239).

the opposite division to the duple proportion; **23** third, the triple proportion mixed out of the duple and sesquialter, and therefore the diapason-plus-diapente; **24** fourth, the sesquitertial proportion, **24a** from which the diatessaron originates **24b** as if it were in the opposite division to the triple proportion; **25** fifth, the quadruple proportion, comprising the bisdiapason, which is determined by the triple and sesquitertial proportions.

26 Nicomachus likewise wished the order of the consonances to proceed in a different manner, so that unity [the number 1] remained the principle of increment and diminution in proportions; **27** accordingly, he also said that the diapason was the fount of the remaining consonances and arranged the other consonances to follow it in the opposite division. **28** Therefore, let the numbers be arranged naturally, beginning from 1, in this way: 1, 2, 3, 4, 5. **29** Then 2 is double of 1, whose reciprocal is the half of 1. **30** [The number] 3 is 3 times 1, whose reciprocal is one third, **31** 4 is 4 times 1, whose reciprocal is one fourth. **32** Thus, the beginning of increase and decrease lies in simple unity. **33** For this reason, the diapason, which, in fact, is the duple proportion, will be placed foremost in the beginning.

34 The diapason excepted, the remaining consonances will be in the opposite order. **35** The sesquialter proportion is opposite the triple division; **36** the sesquitertial proportion is opposite the quadruple division, which is proved by the following argument. **37** The first sesquialter number is 3, if it is compared to 1; **38** then 3, when compared to 2, is the first sesquialter proportion. **39** Now the same first sesquialter number 3 against 1 was the first triple proportion. **40** Moreover, the same 3 yields a triple proportion with the difference between itself and 2. **41** Therefore, since the sesquialter proportion is placed opposite to the triple proportion by this reasoning, the consonance of the diapente is believed to be rightly placed opposite to the consonance of the diapason-plus-diapente. **42** In the same way, the quadruple proportion has the division opposite the sesquitertial, for the first quadruple number, 4, related to unity, is 4:1. **43** It also is the first sesquitertial ratio, if compared with 3; **44** and furthermore, the number 4 is made of its differences with 1 and 3. **45** Thus the sesquitertial [proportion], which is the diatessaron, divides in the contrary manner to the quadruple proportion, which is the bisdiapason. **46** The duple [proportion] has no proportion opposite to it, according to this reasoning of Nicomachus, **47** nor does there exist any number smaller than a sesquialter to

26–28 Boet. *De inst. mus.* 2.20.251.16–28. Concerning the classification of the consonances, Gaf. adheres to the order proposed by Eubulides and Hippasus (cf. Boet. *De inst. mus.* 2.19.251.12–13), not to that of Nicom. (which is also reported here through Boet. *De inst. mus.* 2.18.250.24–25).

29–49 Boet. *De inst. mus.* 2.20.252.1–31, 253.1–8.

which the 2, which is the first duple, can be joined in a superparticular proportion, such that it exceeds the form of the opposite proportion. **48** For that reason also, according to Nicomachus, to the diapason belongs the origin of the consonances. **49** Although these consonances have these relations to each other, we nevertheless are inclined to conclude with the Pythagoreans that it is more proper and better that all multiple consonances precede, and the superparticulars follow.

Chapter **Six**
The consonance of the diapason can only come from the multiple genus

1 There remains [for us] to prove through clearer calculations and more forceful arguments that the consonance of the diapason is formed by the multiple genus. **2** It is known that if we take away from any superparticular proportion another superparticular proportion which is continuous with it—that is, which is smaller—**3** then the [part] left over will be smaller than the mean of the proportion removed. **4** If we remove from a sesquialter proportion a sesquitertial **4a** (which is smaller and successive to it), **4b** the sesquioctaval proportion alone will remain, **5** which falls short of the true half of the removed proportion, that is, the sesquitertial proportion. **6** Thus if we remove a diatessaron from a diapente, a tone remains which does not reach the half of the removed consonance of the diatessaron, **7** because the diatessaron itself is completed by two tones and a smaller semitone.

8 As evidence of this, let us arrange three numbers, namely the sesquitertial and the sesquioctaval proportions, which are contained in the sesquialter, as in figure 4.6.1.

9 You see, for example, how [the ratio] 9:6 brings together the sesquialter proportion and hence the consonance of the diapente **10** and 8:6 gathers the sesquitertial and hence the diatessaron; **11** further, 9:8 makes up the sesquioctaval and for that reason the tone. **12** Therefore, if we remove the smaller number, that is, 6, whose partner in the 4:3 proportion is the middle term, 8, **13** 8 and 9 remain, which against one another produce the sesquioctaval proportion and consequently the tone, **14** by which the true middle of the consonance of the diatessaron cannot be reached. **15** Surely if we remove 5:4 from 4:3, what remains will not reach the middle of 5:4. **16** The same [holds

1 Boet. *De inst. mus.* 2.21.253.28–30.
2–4b Boet. *De inst. mus.* 2.21.254.4–10 paraphrased.
6 Cf. above 4.4.11.
7 Cf. above 4.4.19.
16 Boet. *De inst. mus.* 2.20.254.15–17.

Figure 4.6.1

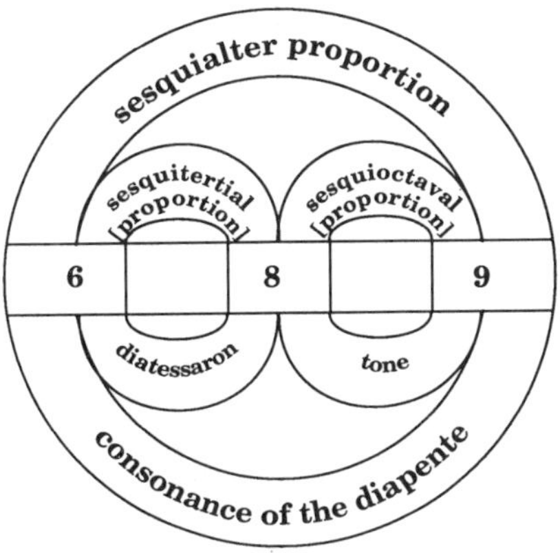

true] for the others **16a** as in these numbers which have already been disposed of.

17 The consonance of the diapason must be taken up in this discussion, **18** because if it is not considered in the multiple genus, **18a** it must be included in the superparticular. **19** Therefore, let this diapason be included as a superparticular proportion. **20** Let us remove the component consonance, that is, the diapente, from it; **21** there then remains a diatessaron, which when doubled surpasses the consonance of the diapente by a tone and a semitone. **22** Moreover, if this diatessaron did not fill half of the diapente, the diatessaron would not be able to fill the diapente when it is doubled. **23** And the diapason, for that reason, may be assigned to a superparticular proportion—**24** certainly because the diatessaron is much larger than half of the diapente. **25** The diatessaron doubled exceeds, as was said, the diapente, **26** for while the diapente consists of three tones and a semitone, the diatessaron is filled up by two tones and a semitone. **27** This diatessaron doubled gathers together four tones and two semitones; **28** in fact, the four tones and two smaller semitones exceed the diapente by a tone and a semitone. **29** Therefore, it is not possible that the diatessaron, taken twice, **29a** should vanish by the subtraction of a whole diapente. **30** For this reason, the consonance of the diapason can by no means be included in a superparticular proportion.

17 Boet. *De inst. mus.* 2.22.254.20 altered.
18 Boet. *De inst. mus.* 2.22.254.21.
18a–23 Boet. *De inst. mus.* 2.22.254.15–30 paraphrased.
24–30 Gaf.'s summary of Boet. *De inst. mus.* 2.22.
25 Cf. above 4.6.21.
26 See fig. 4.4.1.

Chapter **Seven**
To which genus the diapente, diatessaron, and tone belong

1 Now it seems appropriate to demonstrate that the diapente, the diatessaron, and the tone, as previously proposed, are assigned to a superparticular proportion. **2** Anyone who denies this necessarily acknowledges that these intervals belong to the multiple genus. **3** As has been more generally touched on previously, no consonance can be derived from the superpartient genus and from the remaining two composite genera. **4** If, under resulting circumstances, from a whole superparticular proportion a smaller proportion joined to it is taken away, **5** what remains falls short of the middle of this larger proportion, as was concluded and explained in the preceding chapter.

6 Once this principle has been assumed and accepted as truthful, **6a** we try to derive these consonances from the multiple genus. **7** Because the consonance of the diatessaron is smaller, whereas the diapente is larger, the diatessaron occurs in the duple proportion and the diapente in the triple proportion. **8** It is reasonable that if the diatessaron is established in the duple proportion, **9** the diapente, which succeeds it, must be established as a triple proportion, which is successive to the duple. **10** Moreover, the tone, which according to musical custom is in the position after the diatessaron (as the tone is smaller than the diatessaron), will be placed in proportion smaller than the duple proportion; **11** but, in fact this is not found in the multiple genus, for the duple proportion is the first and is smaller than other multiple proportions. **12** Therefore, the tone must be assigned to the superparticular genus. **13** Let the tone, then, be included in the first proportion of the superparticular genus, that is, in the sesquialter proportion.

14 Whenever we take away the duple proportion from the triple, the sesquialter is left. **15** Thus, if the diatessaron is assigned to the duple proportion, and the diapente to the triple, **16** the tone remains after the removal of the diatessaron from the diapente. **17** There is no doubt that the tone must be established in the sesquialter proportion in accordance with this reasoning. **18** But since the tone that is left over cannot fill half the diatessaron—which has been removed from the diapente, in accordance with the reasonable conclusion expressed previously—**19** the two sesquialter proportions now exceed

1 Boet. *De inst. mus.* 2.23.255.3–4.
3 Boet. *De inst. mus.* 2.23.255.10–12 paraphrased; see also above 4.1–2.
4 Boet. *De inst. mus.* 2.23.255.5–8 abbreviated.
5 Cf. above 4.6.14.
6a–7 Boet. *De inst. mus.* 2.23.255.13.
18 Boet. *De inst. mus.* 2.23.255.26–29 paraphrased and abbreviated; see also above 4.6.6.
19 Boet. *De inst. mus.* 2.23.255.29–30; see above 3.8.27.

the duple proportion, as established more generally above. **20** Similarly, the two tones surpass the diatessaron, which is unsuitable, **21** for the diatessaron comprises two tones and a semitone. **22** On the other hand, another option is to assign the tone to the multiple genus, since it is smaller than the diatessaron and, likewise, the diatessaron is smaller than the diapente. **23** Let us assign the tone, then, to the duple proportion, the diatessaron to the triple, and the diapente to the quadruple. **24** Since the diapente consists of a diatessaron and tone, in this manner the quadruple proportion results from the duple and triple, which the order of proportions opposes violently, **25** as the duple and triple proportions naturally produce the sextuple, as illustrated in these numbers: 2, 4, 12. **26** For [the ratio] 4:2 causes the duple proportion, **27** 12:4 the triple, **28** 12:2 the sextuple. **29** If the diatessaron is assigned to the sesquialter proportion, it is proper that the tone, which is smaller, be assigned to the sesquialter, successive to the sesquitertial. **30** But two sesquitertial proportions exceed one sesquialter and, hence, two tones would exceed the consonance of the diatessaron—a fact that musical reasoning denies. **31** From what has been said it is evident that the diatessaron, the diapente, and the tone cannot be assigned to the multiple genus. **32** Thus they belong to the superparticular genus through natural derivation, **33** as all the musical consonances remain part in the multiple genus and part in the superparticular. **34** Likewise, it is necessary that the diapente and the diatessaron adhere to the larger superparticular proportions, **35** which are, of course, the first in the natural arrangement of its genus, that is, the sesquialter proportion and the sesquitertial proportion, which is shown through this proof. **36** If the diapente and the diatessaron belong to the superparticular proportions smaller **36a** than the sesquialter and the sesquitertial, **37** then there is no doubt that just as any other two successive superparticular proportions (apart from the sesquialter and sesquitertial) joined together, would not yield one duple; **38** then diapente and diatessaron together could not fill the consonance of the diapason, which rises from the duple proportion, as has been shown more generally. **39** The duple proportion, too, is composed of the sesquialter proportion and of the sesquitertial. **40** Accordingly, the diapason is made up of the diatessaron and diapente joined together; **41** therefore, if the duple proportion assembles the

20 Boet. *De inst. mus.* 2.23.255.32–33.
21 Boet. *De inst. mus.* 2.23.255.33, 256.1; see also fig. 4.3.1.
22–23 Boet. *De inst. mus.* 2.23.256.3–6 glossed.
24–79 Gaf.'s expansion of Boet. *De inst. mus.* 2.23.256.6–31, 257.1–6; 2.24.257.9–30; 2.25.258.3–4.
24 Cf. fig. 4.2.2.
26 Cf. above 3.3.16–19; 3.4.3; 4.5.17.
27 Cf. above 3.3.20–23a; 3.4.4.

consonance of the diapason, **42** it is necessary that the diapente and the diatessaron are arranged in the sesquialter and in the sesquitertial proportions, respectively, **43** for in no other way can they, joined, fill the consonance of the diapason.

44 If you tried to unite any two other successive superparticular proportions, **45** you would recognize that this duple proportion cannot be completed in any other way, **46** which is illustrated in these numbers: 3, 4, 5, 6, 7, 8. **47** For 4:3 yields the sesquitertial proportion **48** and 5:4 the sesquiquartal. **49** Those proportions united will not complete the duple proportion, since 5:3 cannot be held in the duple proportion. **50** Similarly, the sesquiquintal proportion, joined to the sesquiquartal, is far from the total sum of this duple proportion; **51** for 5:4 makes the sesquiquartal, **52** 6:5 the sesquiquintal, **53** and 6:4 least of all encloses the duple proportion, as it is proven to reach only the sesquialter. **54** This judgment covers also the remaining successive superparticular proportions.

55 It must be noted **56** that any two successive superparticular proportions, if joined, obviously fail to attain the perfection of the duple proportion, for when they are arranged as individual numbers **57** the larger term will fall short of the number necessary for a duple proportion with respect to the smaller term, **58** as in these "prime" numbers: 3, 4, 5, 6, 7, 8, 9, 10. **59** The sesquialter proportion is not subject to this rule **60** since, when the sesquitertial proportion, which is adjacent to it, is joined to it, the duple proportion results in the extreme terms. **61** But beginning from the sesquitertial proportion, **62** to which the sesquiquartal proportion is joined, the combination does not show the duple, **63** since the larger term, that is, 5, falls short by 1 **64** of the number [6]—required to make the duple against the smaller, that is, 3. **65** When the sesquiquintal proportion is combined with the sesquiquartal, they, too, do not produce the duple, **66** for the larger term, that is, 6, is short 2 of completing the duple with the smaller, that is, 4. **67** Moreover, bringing the sesquisextal proportion and the sesquiquintal together will reveal a larger term, that is, 7, which is 3 short **68** of forming a duple with the smaller, 4. **69** If we join together the sesquiseptimal and the sesquisextal proportions, the larger number, that is, 8, will be smaller by 4 than the duple of the smaller number, which

47 Cf. above 3.3.28, 49, 55a–56; 3.4.11; 4.5.16.
48 Cf. above 3.3.30, 61–63; 3.4.12.
51 Cf. above 4.7.48.
52 Cf. above 3.3.30, 50; 3.4.12.
59 : $\frac{3}{2} \times \frac{4}{3} = \frac{2}{1}$
60 That is, 2:3:4 produces 2:4 = 1:2 in the extreme terms.
61 That is, 3:4:5 produces $3{:}5 \neq 1{:}2$.
66 That is, $4{:}5{:}6 \Rightarrow 4{:}6 \neq 4{:}8 = 1{:}2$.

is 6. **70** If we progress in the same order according to the natural arrangement of the numbers, the number by which the larger term falls short of the duple of the smaller term is multiplied for two successive superparticular proportions, **71** as is evident to a diligent inquirer. **72** This happens, because, as has been explained above in [Book] three, the larger superparticular proportions issue from smaller numbers, **73** and the smaller superparticular proportions from larger numbers. **74** Since the sesquialter and the sesquitertial proportions are derived from smaller numbers, they are larger than the remaining superparticulars, **75** and thus permits them to fill out the duple proportion.

76 But the sesquiquartal and the sesquitertial proportions, which contain larger numbers, since they differ by 1, are deficient by 1 for the duple. **77** Since the extremes of the sesquiquartal and sesquiquintal proportions joined differ from the sesquitertial proportion by 2, they are deficient by the same 2 from forming a duple proportion. **78** For the same reason, the sesquitertial and sesquiquartal proportions are larger than the sesquiquartal and sesquiquintal, **79** because larger numbers always yield smaller superparticular proportions, whereas smaller numbers yield the larger superparticular proportion. **80** Thus, since the diapente and the diatessaron are contained in the sesquialter and in the sesquitertial proportions, respectively, as is obvious, **81** and since the sesquialter proportion is larger, but the sesquitertial smaller, **82** and similarly, since the diapente is larger among the consonances, the diatessaron smaller, **83** one must rightly conclude that the larger consonance must be ascribed to the larger proportion and the smaller consonance to the smaller proportion. **84** Consequently, the diatessaron will be assigned the sesquialter proportion and the diapente the sesquitertial. **85** But, for example, if we remove the consonance of the diatessaron from the diapente, that interval which is called a tone, will be left over. **86** Thus whenever we also remove the sesquitertial proportion from the sesquialter, the sesquoctaval will be left over. **87** Therefore, the tone must be constituted in the sesquioctaval proportion.

Chapter **Eight**
The consonances of the diapason-plus-diapente and bisdiapason and of other concords in notes and intervals

1 How the duple and the sesquialter proportions joined yield the triple proportion must now be considered. **2** When the diapason and diapente are

72 Cf. above 3.3.48ff.; 3.4.10ff.
78 Cf. above 4.7.72.
80–87 Boet. *De inst. mus.* 2.25.258.5–16.
1–2 Boet. *De inst. mus.* 2.26.258.19–22.

joined, this will necessarily produce another consonance named by both words. **3** What we call diapason-plus-diapente is naturally derived from the triple proportion. **4** Yet if a sesquitertial is added to a triple proportion, the quadruple will result; **5** here if we add a concord of the diatessaron to a consonance of the diapason-plus-diapente, the consonance of the bisdiapason, which results from this calculation and rests in no other than in the quadruple proportion, will ensue. **6** Also for that reason there are three consonances in the multiple genus, the diapason, the diapason-plus-diapente, and the bisdiapason. **7** It has been proved that the remaining three consonances are reasonably well established in the superparticular genus, that is, the diapente, the diatessaron, and the tone.

8 When these consonances are considered critically, and consonance is thought of as the mixing of two notes, **8a** according to Boethius, **8b** then consonance can in no way be made without tone. **9** Tone is the incidence of a moderated voice produced in one tension. **10** Therefore what is judged to be the smallest particle of melody (i.e., pitch), results from vibration. **11** And every vibration arises from motion, as has been described more thoroughly in [Book] two.

12 Some motions are equal, others are unequal. **13** Some of the unequal motions are more unequal than others. **14** Some, indeed, are moderately unequal. **15** Consequently, the equality of pitches must be assumed to reside in the equality of the motions. **16** From a moderate inequality that is rather evident result the nearest and the most natural proportions, that is, the multiple and superparticular, **17** among which are the duple, the triple, the quadruple, the sesquialter, the sesquitertial, and the sesquioctaval. **18** From the various unequal proportions, which are either not so clear or whose terms stand far apart from each other, the dissonances emerge, **19** for no concord is produced from these pitches. **20** Thus, the Pythagoreans did not attempt to consider any consonance beyond the bisdiapason, because of the greatest inconvenience and distance of these most remote pitches.

21 If anyone were to advance to the more remote intervals, **22** he would experience that the same disposition of the tones or semitones proceeds naturally as we have learned art and nature observe, in the rest of the conso-

3–5 Boet. *De inst. mus.* 2.26.258.22–27.
6 Boet. *De inst. mus.* 2.20.252.31, 253.1–5 glossed.
7 Boet. *De inst. mus.* 2.20.253.8 expanded.
8, 9–11, 20 Boet. *De inst. mus.* 2.21.253.9–25.
9 Aristox. *Harm.* 1.15. For a translation of this passage by various authors, both ancient (e.g., Antonio Gogava) and modern (e.g., Henry S. Macran), see Palisca, *Humanism,* p. 150.
11 Cf. above 2.1.28ff.
20 Cf. above 4.4.39.
20–31 Gaf. is largely indebted to Boet. *De inst. mus.* 2.27.

nances arranged above. **23** But the more the pitches are stretched beyond the bisdiapason, the more evident it becomes that dissonance is brought to the ears by the distance of the most remote pitches. **24** I persuade myself that this occurs, because the most remote pitches of the same notes are judged in this more abundant distance to depart from their equalness. **25** These pitches are easily examined by the judgment of the ears and by the natural arrangement of the multiple proportions. **26** For that reason, in investigating the consonances, Boethius and the other musicians have invented the fourfold space, fixing sufficiently distant appropriate extremities, **27** for which reason the musical consonances ought not to be extended beyond these limits by any further stretching.

28 If it is necessary that these consonances, having been derived from different proportions, comprise manifold pitches, **29** it will be permitted also to consider these consonances by different intervals in a reasonable measure; **30** and for this reason, the following treatise arranges these pitches and also the intervals according to a clearer resonance. **31** The [whole] tone is made from two steps and one interval. **32** The diatessaron is completed by four steps and three intervals. **33** The diapente comprises five steps and four intervals. **34** The diapason, derived from the diapente and the diatessaron, is filled by eight steps and seven intervals. **35** The diapason-plus-diapente, embracing twelve steps, includes eleven intervals. **36** The bisdiapason, which, finally, holds fifteen steps, retains fourteen intervals. **37** Accordingly, each consonance always has one interval less than it has notes. **38** It should also be known that the diapente [h1r] and the diatessaron conjoined, yield the diapason. **39** If separated, they would exceed this diapason by a single whole tone and by an interval. **40** Therefore, when we speak of compound consonances, we understand that one step is common, that is, the higher step of the lower consonance is the lower step of the higher consonance.

32 Boet. *De inst. mus.* 1.17.203.18–19.
33 Boet. *De inst. mus.* 1.18.204.11–12.

BOOK **FIVE**

Chapter **One**
The invention and arrangement of the tones

1 Nicomachus reports that the practice of singing was first so simple that it consisted in only one tetrachord, **2** whose first step answered the most remote step in a consonance of the diapason, and the first step concorded with the third step by a diapente, and also the third step with the fourth step by a diatessaron, and the second step to the third step yielded a whole tone. **3** The Greeks assert that Mercury was the inventor of this tetrachord. **4** But I believe rather that Jubal or one of the first men prior to the Flood brought together tones in some way or another. **5** By means of these four tones, the practice of singing, arranged from the four elements, seemed to imitate the music of the spheres. **6** Furthermore, singing soon after was guided by a good many people and little by little was increased in many ways. **7** For Chorebus, son of Athiris, who was king of the Lydians **7a** (if I may trust the Greeks themselves), **7b** added the fifth step on top. **8** Then, Hyagnis, who was a Phrygian, **8a** added the sixth step. **9** It was Terpander, from Lesbos, who added the seventh step. **10** He was led by the order of the seven planets **11** and, with nature as his guide, he decided that the seven varied and dissimilar tones would be produced by these seven steps one after the other; **12** and they are dissimilar, because one cannot be compared with another by any sort of similitude. **13** In fact, these tones exhaust all varieties of harmony, once replicates have been accounted for. **14** Virgil appears to have celebrated these [tones] in the sixth [book of the *Aeneid*], writing, **15** "they

1–34 D'Abano *Expo. Pr. Arist.* 19.3.
1–2 Boet. *De inst. mus.* 1.20.205.28, 206.1–6; Nicom. lost *opus maior* (cf. Bower, trans., *Boethius: Fundamentals,* p. 29); Cens. *Fragm.* 12; Pliny *HN* 7.56.204; Strab. *Geogr.* 13.3–4.
3 Boet. *De inst. mus.* 1.20.206.6–7; see also Gaf. *De harm.* 1.1, 1.4. Concerning sources transmitting Mercury's invention of the lyre, see Bower, trans., *Boethius: Fundamentals,* p. 30.
7, 7b Boet. *De inst. mus.* 1.20.206.8; Ans. *De mus.* 1.122; Plut. *De mus.* 1136C. Concerning the correct reading ("Toroebus") in Boet. *De inst. mus.*, see: Bower, trans., *Boethius: Fundamentals,* p. 3, and Wille, *Musica Romana,* pp. 670–671.
8 Boet. *De inst. mus.* 1.20.206.9; Ans. *De mus.* 1.122; Ath. *Deip.* 14.624.
8a Boet. *De inst. mus.* 1.20.206.9–10; Ath. *Deip.* 14.624.
9–10 Boet. *De inst. mus.* 1.20.206.10–11. Further on Terpander's association with the heptachord lyre through three distinct traditions, see Bower, trans., *Boethius: Fundamentals,* p. 31.
14–15 Virg. *Aen.* 6.646; Guido *Microl.* 5.

spoke out the tuneful divisions of pitches in numbers," that is, seven different pitches. **16** This is a metaphor of the seven planets producing the music of the spheres and the seven days of the week. **17** Their whirling motion multiplied many times are recognized to complete the entire body of the year. **18** Indeed, this progression by sevens appears to have an incredible power and ability in many things of nature. **19** Varro discussed the number 7 at great length in his *Hebdomades vel de imaginibus*]. **20** Of these seven steps, the lowest is named hypate, as if it were greater and more honorable. **21** Hence also, they call Jupiter hypaton, that is to say, "principal"; **22** and likewise, they call the consul by the same name because of the excellence of his dignity. **23** This property of the hypate is in fact attributed to Saturn as a result of the slowness of its motion and the depth of its pitch. **24** The second step is designated parhypate, since it is fixed next to the hypate. **25** The third step is designated lichanos, because the finger which we call index finger is named lichanos by the Greeks, **25a** derived from their word for "licking." **26** In accompanying singing, this finger should reach for this step on the lyre, which is the third from the hypate; so it is called lichanos by musicians and for that reason retains the name [h1v]. **27** The fourth step is called mese, that is, the middle, because four is always the midpoint between one and seven. **28** The fifth step is called paramese, since it is placed next to the middle note. **29** The seventh step is called nete, that is, neate, which means inferior. **30** In between this nete and paramese, the sixth step appears, which is called paranete, because it is placed next to the nete. **31** Since the paramese is the third step from nete, the paramese is called by the very name of trite, that is, the third. **32** We arrange all these seven steps in this order in a diagram (fig. 5.1.1).

33 These seven steps are deemed to be quite essential and altogether necessary, since they are, as has been stated, the only ones different among themselves and from which all melody is derived. **34** Pietro d'Abano also declares this most clearly in his elucidation **34a** of the third problem of Aristotle's Music [Problem 19.3 of pseudo-Aristotle]. **35** When we climb beyond these seven tones in singing a melody, a tone appears that is similar to the first lower step, **36** and if we proceed beyond, a tone similar to the second

16–20 Varro *Heb.* (lost), a work in twenty books on famous Romans and Greeks, including 700 portraits (cf. Pliny *HN* 35.11). In *De harm.* 1.4 Gaf. names the lowest step the proslambanomenos, likewise the second step hypate hypaton, and so on. On the Greek designation of steps, see F. Zaminer, "Hypate, Mese und Nete im frühgriechischen Denken: Ein altes musikterminologisches Problem in neuem Licht," *Archiv für Musikwissenschaft* 41 (1984):1–26.

21–22 Boet. *De inst. mus.* 1.20.206.13–15.

23 Boet. *De inst. mus.* 1.20.206.15–16. Concerning the association of the hypate with Saturn, see Boet. *De inst. mus.* 1.27.

24 Boet. *De. inst. mus.* 1.20.206.16–18.

25 Boet. *De inst. mus.* 1.20.206.18–19; Huc. *De mus.* (cf. Gerbert, *Scriptores* 1:117).

25a–32 Boet. *De inst. mus.* 1.20.206.19–29, 207.1–7.

Figure 5.1.1

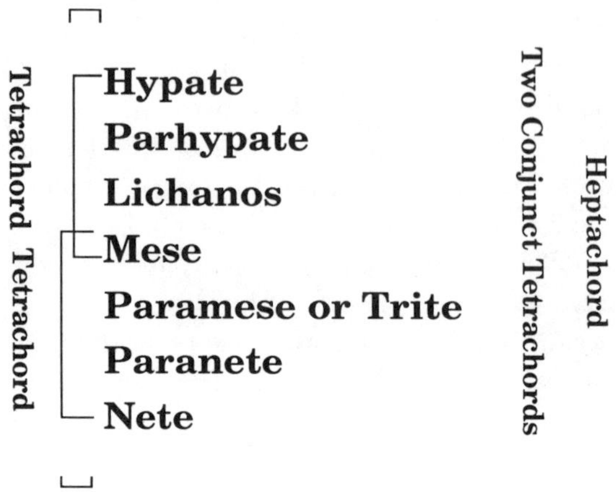

step occurs; **37** and in this manner, the seven different tones are duplicated. **38** Accordingly, the eighth tone is similar to the first, the ninth to the second, the tenth to the third, the eleventh to the fourth, the twelfth to the fifth, the thirteenth to the sixth, the fourteenth to the seventh; **39** the fifteenth replicates the eighth, obviously, since it resembles the first tone, too. **40** Therefore, only these very seven tones are different and dissimilar, **40a** as elsewhere Virgil expresses it, **40b** "a pipe, fitted together from seven stalks of hemlock."

41 The eighth tone is the first that resembles the first. **42** Any sound, by nature, is produced similar to its eighth; **43** for that reason, the consonance of the diapason accomplishes such an arrangement of the most remote tones, **44** so that, according to Ptolemy, one and the same tone seems to be present, **45** which is agreeable to Plato **45a** and to the Pythagoreans also; **46** [this arrangement] is completed by eight tones.

47 When this disposition of the seven steps had been established, Lycaon, from Samos, added an eighth step, **48** in order to complete the consonance of the diapason, which resides in the octachord. **49** This [eighth step], he inserted between the paramese, also called trite, and the paranete, so that it was the third from the nete. **50** And this step was called paramese, when it alone is placed after the middle tone. **51** Thus it lost its name of trite, since between it

40b Virg. *Ecl.* 2.36–37.
44 Boet. *De inst. mus.* 5.19.358.23–24; Ptol. *Harm.* 1.6.
45 Pl. *Resp.* 670a.1.6–671a.1.1.
45a Boet. *De inst. mus.* 5.9.358.24–25.
47–52 Boet. *De inst. mus.* 1.20.207.8–25. On the interpretation of Lycaon of Samos as a veiled reference to Pythagoras of Samos, see Bower, trans., *Boethius: Fundamentals,* pp. 32–33. Nicom. (*Man.* 5.1) attributes the invention of the eighth step to Pythagoras.

Figure 5.1.2

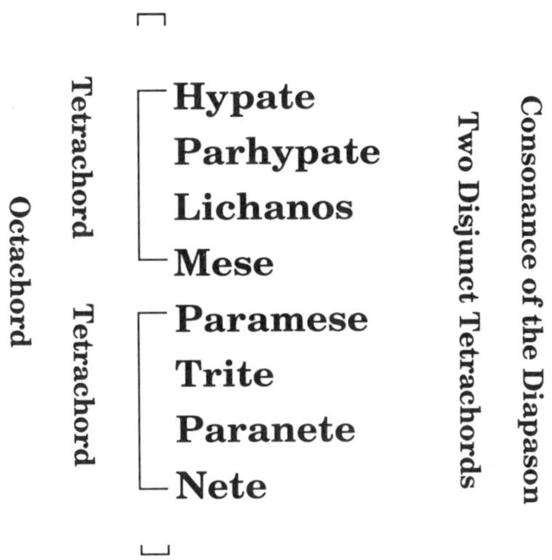

and the paranete a third step was placed from the nete, and it not unjustly claimed the name trite. **52** This then is the arrangement of [Lycaon's] octachord [h2r] (fig. 5.1.2).

53 Two tetrachords are joined in a heptachord. **54** Moreover, the synaphe, which in Latin is called "conjunction," occurs in this heptachord whenever the mediating function of a bounding note is to join together and connect two tetrachords; **55** and, in fact, in the above representation of the heptachord, the mese, which is the middle note, is common to these two tetrachords; **56** for the last step of the first tetrachord is the first step of the second tetrachord. **57** Therefore the heptachord is called synemmenon, that is, "joined together"; **58** the octachord, however, is called diezeugmenon, because it is disjunct; **59** for in an octachord two tetrachords are separated because the first four steps complete the first, while the remaining four complete the second tetrachord. **60** In an octachord, therefore, there exists a diezeuxis, which can be called disjunction, as often as two tetrachords are separated by the mediating function or by an interjection of a whole tone or a semitone. **61** Indeed, when there is a mese, which takes its name from the middle position, it is not properly the

54 Boet. *De inst. mus.* 1.24.217.4–8.
57–58 Boet. *De inst. mus.* 1.20.207.25–27. In *Theorica* and *Theoricum* "diezeugmenon" (Boet. *De inst. mus.* 1.20.207.26) is replaced by "mezemeinon"—obviously a copying error.
60 Boet. *De inst. mus.* 1.20.208.5–8 paraphrased; cf. also Boet. *De inst. mus.* 1.25.
61–64 Boet. *De inst. mus.* 1.20.208.8–13. In *Theorica* and *Theoricum* Gaf. replaces Boet.'s "Prophrastus" (cf. *De inst. mus.* 1.20.208.10) with "Theophrastus." Further on Prophrastus, see Bower, trans., *Boethius: Fundamentals,* p. 34; also Wille, *Musica Romana,* pp. 671–672.

middle note in the octachord; **62** on the contrary, there are two midpoints [i.e., two notes at the middle]. **63** Theophrastus the peripatetic added one step to the lower part [of the octachord]; **64** because it was placed above the hypate it was called parhypate. **65** The musically gifted Estiachus added the tenth step in the lower part [of the octachord]. **66** Timotheus from Milet installed an eleventh step. **67** Since these were added beyond the hypate and the parhypate, the first of them was called hypate hypaton, that is, the greatest of the larger steps and the lowest of the low. **68** So, of these eleven steps, the first one is called hypate hypaton **69** and the second parhypate hypaton, that is, placed next to the hypate hypaton. **70** The third step, previously called lichanos in the heptachord, was named lichanos hypaton. **71** The fourth retained the ancient name hypate. **72** The fifth is parhypate, the sixth lichanos, the seventh mese, the eighth paramese, the ninth trite, the tenth paranete, and the eleventh nete. **73** There is in this endecachord, in fact, one tetrachord of the form hypate hypaton, parhypate hypaton, lichanos hypaton, and hypate, **74** and another of the form hypate, parhypate, lichanos, and mese; **75** and those two tetrachords are indeed conjunct, **76** for the hypate is the higher step of the first and lower tetrachord but also the lower of the second, higher tetrachord. **77** Indeed, the third tetrachord is composed of paramese, trite, paranete, and nete. **78** Between this first tetrachord and the third is placed a middle tetrachord, that is to say, the second tetrachord of the form hypate, parhypate, lichanos, and mese; therefore this whole tetrachord is called meson, that is to say, "of the middle," **79** and its steps are named with the addition of the word "meson," that is, hypate meson, parhypate meson, lichanos meson, and mese. **80** Since between the second tetrachord and the third the mese is separated by a tone from the paramese, this third tetrachord is called the tetrachord [h2v] of the "disjunct," **81** and its steps are named with the addition of the word "diezeugmenon" in this manner: paramese diezeugmenon, trite diezeugmenon, paranete diezeugmenon, and nete diezeugmenon; **82** in fact, this description of the endecachord brings together in a very clear manner the disposition of all these steps of the three tetrachords (cf. fig. 5.1.3).

83 Thus there is a disjunction between mese and paramese, and, therefore, this third tetrachord is called diezeugmenon. **84** Truly, if one takes away the paramese and also the diezeugmenon tetrachord, the following tetrachord remains of the form mese, trite, paranete, and nete; **85** then there will be two

65 Boet. *De inst. mus.* 1.20.208.29, 209.1–30, 210.1–17. In *Theorica* and *Theoricum* Gaf. replaces Boet.'s "Histiaeus of Colophon" (cf. Boet. *De inst. mus.* 1.20.208.29) by "Estiachus." On Histiaeus, see Bower, trans., *Boethius: Fundamentals,* p. 35.

66–85 Boet. *De inst. mus.* 1.20.209.1–30, 210.1–17.

70 Gaf. erroneously here labels the third step of the heptachord "parhypate." This translation and edition have been emended to read "lichanos" (cf. also fig. 5.1.1).

Figure 5.1.3

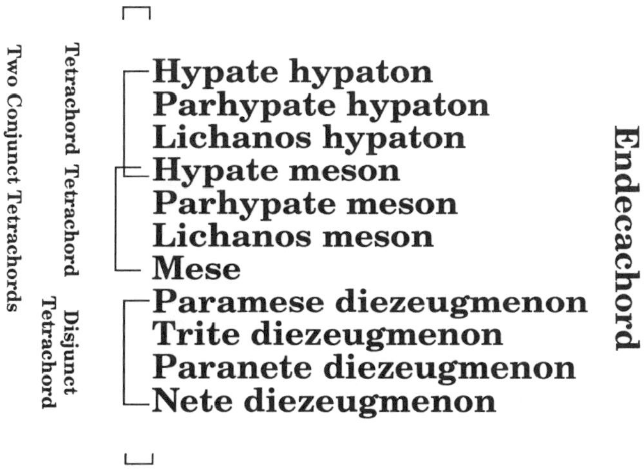

conjunct, that is, synemmenon tetrachords of this type: mese, trite synemmenon, paranete synemmenon, and nete synemmenon. **86** In fact, the synemmenon tetrachord is added and is united with this step [i.e., the mese] in order to soften the harshness of the tritone; **87** art renounces and nature shudders at this dissonant and harsh interval. **88** In fact, this tetrachord would be superfluous and rejected if the dissonant and harsh tritone itself, placed between the parhypate meson and the paramese, could adapt itself in its natural character of sound to the disposition of the diatonic scale. **89** For this reason, only the first three tetrachords, that is, the hypaton, meson, and diezeugmenon, were recognized by the philosophers, in the natural division of the monochord and were so arranged in the most often repeated observation. **90** But because the mese, which derives its name from its middle position, was not the middle note in the endecachord, another tetrachord was joined above the nete diezeugmenon. **91** Since this tetrachord exceeds the remaining steps by the acuteness of its tones, it was called hyperbolaion, that is to say, "added to the top." **92** And this tetrachord is conjunct to the nete diezeugmenon step in this order: nete diezeugmenon, trite hyperbolaion, paranete hyperbolaion, and nete hyperbolaion. **93** In this arrangement, four tetrachords contain the fourteen steps; **94** the first two are conjunct, and the following two are in turn conjunct, but these last two tetrachords are disjunct from the first two tetrachords. **95** These tetrachords are arranged accordingly: since the mese did not remain in the middle position, a single step was added to the lower part, called

87 This is presumably a reference to Anon. *Mus. ench.* 12.
90–92 Boet. *De inst. mus.* 1.20.210.28–29, 211.1–20.

Figure 5.1.4

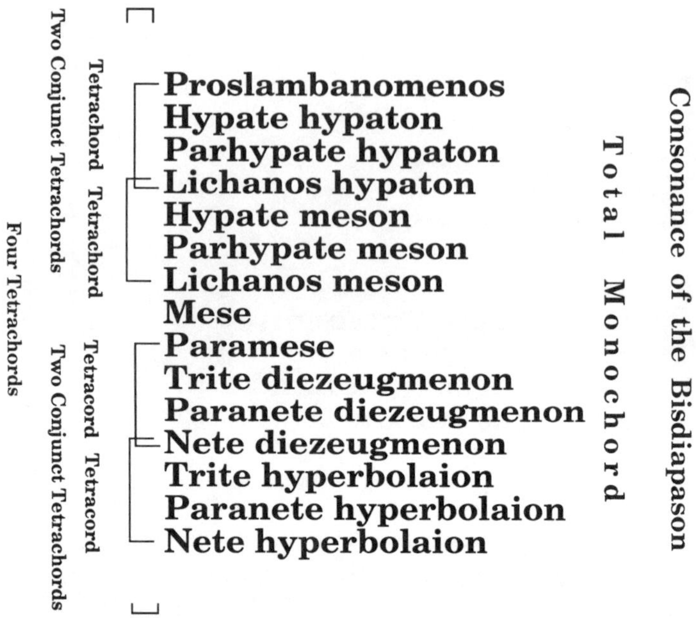

proslambanomenos, that is, "acquired" or "added." **96** Although it was placed at the far end of the scheme, it was necessary to put it before the others. **97** Consequently, the philosophers themselves maintained the monochord in the fifteen steps, through which the consonance of the bisdiapason is brought together by a marvelous logic. **98** Thus this arrangement was extended [h3r] (cf. fig. 5.1.4).

99 Finally, it is agreed that among the Greeks there were these five tetrachords, that is, hypaton, meson, synemmenon, diezeugmenon, and hyperbolaion. **100** Of these, the middle one, that is to say, the synemmenon, was not accepted in the middle of the arrangement of the fifteen steps of the monochord, since this tetrachord seemed to be superfluous; **101** as we have certainly mentioned before, the synemmenon tetrachord was derived in order to soften the harshness of the tritone. **102** We suspect, however, that this tetrachord was not invented for the purpose of eliminating the tritone or suppressing it from the system of the monochord, **103** because, by nature, the tritone will occur in the formation of many species of diapente and diapason, which cannot be brought together under any circumstances without the

Fig. 5.1.4 Boet. *De inst. mus.* 1.21.212.8–22.
99 Boet. *De inst. mus.* 1.25.218.18–20.
101 Cf. above 5.1.86–88.
103 For a detailed discussion of this topic, see below 5.7–8.

tritone itself. **104** Therefore, the synemmenon tetrachord, which adheres to the mese step, is included as one pleases or likewise omitted as one pleases. **105** Moreover, Albinus translated the names of the steps of these five tetrachords into Latin. **106** Namely, he called the hypaton "principales," the meson "mediae," the synemmenon "coniunctae," the diezeugmenon "disiunctae," and the hyperbolaion "excellentes."

107 In every single tetrachord the diatessaron sounds as a consonance. **108** The designation "tetrachordum" is derived from "tetra," which is four, and from "corda"; **109** for this reason, the tetrachord represents a union of four steps. **110** Likewise, "diatessaron" is derived from "dia" which means "de" (from) or according to some people from "dia," that is, "through," and from "tessara," meaning four, that is the consonance is made of four tones. **111** According to the grammarians, "dia" written with the Latin "i" **111a** is the same thing as the preposition "de," but if it is written with a Greek "y" it means "two."

Chapter **Two**
The three genera of tetrachords

1 Tetrachords are classified by philosophers into three genera of melodies. **2** Melos is so called after its honey-sweetness [*mel*], whence melodia, that is, "sweet song." **3** Thus they call the first genus the diatonic in which any tetrachord proceeds by smaller semitone, whole tone and whole tone. [h3v]

4 In this diatonic genus the tetrachord always consists of three intervals: **5** the first, which lies between the first and second steps, is called a smaller semitone; **6** the second, which emerges between the second and third, is a whole tone; **7** and the third, between the third and fourth steps, fills the space of a whole tone. **8** Therefore, I believe that the diatonic genus is so called because it naturally contains a ditone of two intervals. **9** This genus, because it is somewhat harder and more natural, seems to have been esteemed by philosophers more than the others **9a** and is recognized in its frequent use by musicians and, guided by nature, by the discipline itself.

105–106 Boet. *De inst. mus.* 1.25.218.22–24, 219.1–2.
110–111a Similar etymological derivations are found in contemporary treatises, for example, Jac. *Spec. mus.* 5.10. For a brief discussion of this chapter, see F. Joseph Smith, "The Medieval Monochord," *Journal of Musicological Research* 5 (1984):8. The identification of "dia" with "duo" surfaces for the first time in the region of Liège in the eleventh century. On this particular etymology as well as for an overview of similar etymologies, see Fritz Reckow, "Diapason, diocto, octava," *Handwörterbuch* (1976), pp. 9ff.
1–3 Boet. *De inst. mus.* 1.21.212.24–25; 4.7.322.22–26 paraphrased.
4–7 Boet. *De inst. mus.* 1.23.216.18–24 paraphrased.
8–9 Boet. *De inst. mus.* 1.22.212.26–27.

Figure 5.2.1

Tetrachord in Diatonic Genus

HYPATE HYPATON

semitone

PARHYPATE HYPATON

tone

LICHANOS HYPATON

tone

HYPATE MESON

Consonance of the Diatessaron

10 As in all melody-making this modus is more natural, **10a** so that within the single tetrachord **10b** or within the four individual tones, the consonance of the diatessaron, composed of two separate tones and one semitone, always emerges. **11** This agrees very well with the diatonic genus **11a** but not at all with the chromatic or the enharmonic. **12** To be sure, the diatonic tetrachord proceeds by smaller semitone, whole tone, and whole tone, beginning with the semitone, **12a** which is softer and also more pleasant in sound. **13** In fact, philosophers, in investigating the tetrachords, built them up from a rather sweet beginning and proceeded through harsher and harder tones. **14** For this reason, the synemmenon tetrachord would entangle the [para]mese step with the harshness of its tritone—**14a** which would not happen unless these tetrachords began with a semitone step. **15** Therefore, when we make melody through four discrete tones one by one, we naturally produce the consonance of the diatessaron; **16** the diatessaron itself requires a threefold tuning in the diatonic genus. **17** This type of singing is called "triplex." **18** For the first progression of melody or type of singing results when we derive the first species of diatessaron in four steps, that is, reckoned by ascending tension through one semitone and two whole tones and by descending relaxation in the opposite way. **19** This is called the first species of diatessaron. **20** The second type of singing occurs when, with these four contiguous steps of tension that we derived, we sing out in succession two whole tones and one semitone, from which the second species of diatessaron is made. **21** The third method of making melody results when in ascending we bring forth a whole

Fig. 5.2.1 Boet. *De inst. mus.* 1.22.215.20ff.; see also Gaf. *De harm.* 1.5.
12 Boet. *De inst. mus.* 1.23.216.20–21.

tone, then a semitone, and finally another whole tone; and in descending likewise. **22** In the arrangement of these tetrachords this is called the third species of diatessaron. **23** But if we proceed to the more remote steps, we shall observe that when these four steps are sung one by one nothing occurs that is different from these three described species of diatessaron which will be replicated in the same progression. **24** The species of diatessaron differ among themselves solely in the placement of the semitone.

25 This was the disposition of the diatessaron for the different species in the monochord set up for tetrachords alone. **26** But then the proslambanomenos step was added beneath in the lower part, distant by a tone from the hypate hypaton. **27** This addition does not seem useless, for the proslambanomenos step completes a consonance of the diapason to the mese, the middle of the fifteen-step system, a bisdiapason to the farthest step, the nete hyperbolaion, **28** and to the lichanos hypaton it completes that species of diatessaron which is composed of a whole tone, semitone, and whole tone, which is the acknowledged third among the tetrachords. **29** Later musicians numbered the three species of diatessaron from the proslambanomenos step; **30** and for this reason the species which was the third is now the first, proceeding by a whole tone, semitone, and whole tone. **31** The one species which was then first is now second, ascending a semitone and two whole tones. **32** That species which was second is transformed into the third, proceeding by two whole tones and a semitone. **33** In fact, musicians follow and respect these three species of diatessaron so disposed in the composition of the modes. **34** In this connection it should be recalled that there are only three species, given that the diatessaron is composed of four notes, **35** for every consonance always has one species less than it has notes. **36** Therefore, the musicians demonstrated these three species of diatessaron in the above arrangement of the tetrachords, **37** namely, that two tetrachords are joined together in one heptachord, and the first replicated species lies between the fourth and seventh steps. **38** Now, indeed, the first six steps are perceived to contain these species, derived from the proslambanomenos step, **38a** as is evident in this arrangement (cf fig. 5.2.2).

39 Since these three species of diatessaron are produced with six steps only, we run across them repeatedly in the layout of the monochord. **40** The Latin theorists are known to venerate Guido of Arezzo as the institutor of the musical *Introductorium,* **40a** which they call "monochord" **40b** or "hand" **40c** with six syllables, that is, *ut re mi fa sol la* repeated seven times. **41** By these syllables, we easily perceive that these three species of diatessaron can be

35 Cf. also below 5.7.14.
40 Guido's *Introductorium* refers to a compilation of the *Reg.*, the Prologue to the Antiphoner, and the *Epist.* (cf. Claude V. Palisca, "Guido of Arezzo," *New Grove* 7:807).

Figure 5.2.2

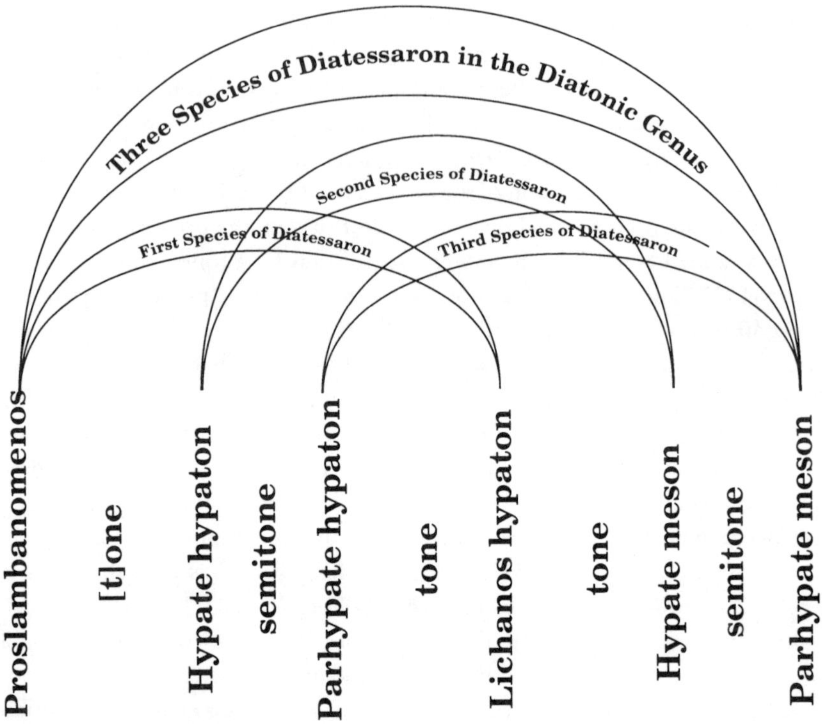

constructed and repeated as often as one wants. **42** Therefore, since the diatonic genus connects the tetrachords appropriately with three distinct intervals through the diatessaron, that is to say, by the semitone, and also whole tone and whole tone, proportionately and consistently derived, **43** it has been retained by musicians and indeed remarked on for its very suitable harmony. **44** For every harmonic consensus is conceived in the genus itself; **45** for it happens that the material is arranged, as if in a more noble and honored progression of the possibilities presented. **46** The other genus in which the tetrachords have been arranged by these philosophers was called the chromatic; **47** the Greek "chroma" in Latin means color; **48** from there stems the name "chromaticum," that is, one produces the tetrachord by a different way of ascending, as if colored. **49** For as one surface changes into another, one color runs into another; **50** so this genus of singing, insofar as the color changes, is different from the rest.

Fig. 5.2.2 Boet. *De inst. mus.* 1.22.215.20ff.
47 Boet. *De inst. mus.* 1.21.213.8–9.
48–52 Boet. *De inst. mus.* 1.21.213.9–19 glossed and abbreviated.

Figure 5.2.3

Tetrachord in Chromatic Genus

HYPATE HYPATON

smaller semitone

PARHYPATE HYPATON

larger semitone

LICHANOS HYPATON

semiditone

HYPATE MESON

Consonance of the Diatessaron

51 This genus forms the ascending tetrachord by means of the smaller semitone, the larger semitone, and three semitones, of which one is larger, that is, the apotome. **52** The apotome and two smaller semitones complete the semiditone, that is the tone plus two smaller semitones. **53** Every single tetrachord in this genus [h4v] contained the consonance of the diatessaron, but not by the same intervals by which it was shown in the diatonic genus. **54** To take an example from the remaining tetrachords, let us first arrange the tetrachord by means of its proper intervals. **55** From the hypate hypaton, they ascended to the parhypate hypaton by means of a smaller semitone. **56** But the lichanos hypaton is a larger semitone higher than the parhypate hypaton; **57** between the hypate meson and the lichanos hypaton were two smaller semitones which together with the apotome completed the semiditone; **58** it was evidently named after "semum," because it was imperfect, and "ditonus," that is, two almost incomplete tones. **59** Figure 5.2.3 is a diagram of this tetrachord.

60 Actually, since the larger semitone is not reckoned among the consonances on account of its disagreeable disposition, **61** theorists most ineptly and inconveniently failed in an attempt to measure the semiditone by the interval of two neighboring steps, with no other step inserted in between, and to use it in melody because it is comprehended with difficulty by the least interval of two contiguous steps, **62** so much the less can it be contained in two intervals of three steps **63** so that anybody could easily perceive it among the consonances by his own observation. **64** Later theorists, having proved this genus to be worthless, utterly abandoned it.

Fig. 5.2.3 Boet. *De inst. mus.* 1.22.215.20ff.; see also Gaf. *De harm.* 2.2.

Figure 5.2.4

Tetrachord in Enharmonic Genus

HYPATE HYPATON

diesis

PARHYPATE HYPATON

diesis

LICHANOS HYPATON

ditone

HYPATE MESON

Consonance of the Diatessaron

65 Another genus was called the enharmonic, as if to say that it filled the tetrachord more suitably and agreeably than the chromatic. **66** In this genus the philosophers divided the smaller semitone into two dieses. **67** The diesis is the half of the smaller semitone, **67a** as was explained more fully in the preceding chapters. **68** Therefore, in this genus the tetrachord ascended from the first step to the second by means of a diesis; **69** likewise, from the second step to the third; from the third step to the fourth by an interval of the ditone, as is evident in this tetrachord (cf. fig. 5.2.4).

70 In each genus, the proslambanomenos preserved its own nature. **71** For it was always one whole tone away from the hypate hypaton; **72** the proslambanomenos was not subjected to the order of these steps of the tetrachords **73** so that when this order was varied according to the measured arrangement of the several genera, it would have been moved. **74** After the two genera were cast aside and left behind, the proslambanomenos step was preserved in the diatonic genus, which alone was retained and most frequently used. **75** The proslambanomenos itself was reckoned together with the remaining steps in the formation of the consonances.

76 In connection with the diatonic genus, no consonance smaller than the smaller semitone was recognized in theory, as will be evident in the following chapters. **77** The diesis, on account of the smallness of its interval, has no sound other than that of the interval of which it is a part. **78** Since the ditone

Fig. 5.2.4 See also Gaf. *De harm.* 2.9.
67 Boet. *De inst. mus.* 1.21.213.17.
67a *Theorica* 4.3.11, 15 (cf. *Theoricum* 5.2.174).
76 *Theorica* 5.3–5.

cannot be comprised in one least interval of two successive steps in a natural way, **79** because it is certainly understood to be completed by the two intervals of three successive [h4v] steps, **80** this enharmonic genus so went out of use due to its too great disagreeableness and difficulty of usage. **81** Therefore, in the individual genera every single tetrachord preserved the consonance of the diatessaron in different **81a** and various ways of ascent. **82** The chromatic and enharmonic genera, however, on account of their softness in performance and their difficulty, were rejected by the musicians. **83** In fact, the Pythagoreans, Platonists, and other musicians were accustomed to make frequent use of the diatonic genus—on account of its well-fitted and naturally arranged derivation—with the greatest enthusiasm.

Chapter **Three**
The intervals of the fifteen steps

1 The *diastemata* or intervals of the fifteen steps arranged above must now of necessity be defined. **2** The proslambanomenos, the first step, is lower than the remaining steps; **3** the second step, on the other hand, is the hypate hypaton, whose name translates as the "lowest one of the low" (which name is perfectly justifiable). **4** Before the proslambanomenos step was added, **5** then, the hypate hypaton was the first and lower step **6** although now, however, it is higher by a whole tone than the first step; **7** the hypate hypaton has retained the name assigned to it at first. **8** The third step is the parhypate hypaton, which goes upward from the hypate hypaton by a mere smaller semitone. **9** The fourth step is the lichanos hypaton, which is one whole tone away from the parhypate hypaton and which with the proslambanomenos step makes a consonance of the diatessaron in the diatonic genus. **10** The fifth step is the hypate meson, which is translated as "lowest of the middle ones" and which is higher by a whole tone than its neighbor which precedes it, **11** and it completes the consonance of the diapente with the first step. **12** The sixth step is the parhypate meson, which departs from the fifth step by the single interval of a smaller semitone. **13** The seventh step is the lichanos meson, which is apart from the sixth step by the ascent of a whole tone. **14** The eighth step is the mese, which means "middle," formed by ascending a whole tone from the seventh step, its neighbor. **15** It completes the diatessaron with the fifth step and the diapente with the fourth step. **16** Also, the eighth step is distant from the first step by the consonance of the diapason, that is to say, through the duple proportion. **17** This step, in fact, resides in the middle position and

2–14 Boet. *De inst. mus.* 1.22.214.4–14.

Figure 5.3.1

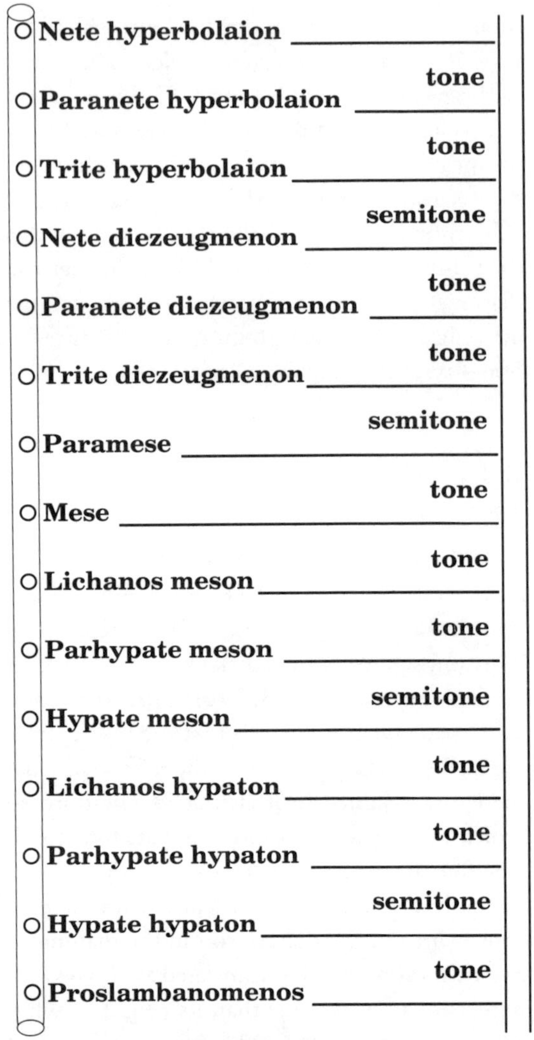

controls the remaining subsequent steps—which are related to the eighth step as if to their beginning and foundation of the relation. **18** The ninth step is the paramese, which is a whole tone away from the mese **18a** and which, as a tone

Fig. 5.3.1 Boet. *De inst. mus.* 1.21.212.8–22.

with two preceding tones, in fact comprises the harsh and severe tritone in the four steps. **19** This step is kept with necessary caution by the philosophers in this position **20** so that this ninth step would complete with the second step—the hypate hypaton, sitting eight places away from it—**20a** one of the seven perfect species of the diapason, with its five tones and two smaller semitones. **21** However, if this paramese step were only one smaller semitone away from the mese step, **21a** then the consonance of the diatessaron, drawn out of the natural diatonic arrangement, would be contained in the four steps themselves. **22** But when the philosophers had researched the nature of things for a rather long time, at the very first a manner of harmony presented itself as the most suitable—**22a** a manner by which they escaped the dreadful unsuitability of the tritone, at which even nature shudders in making melody; **23** so the philosophers preserved this second diapason and also the complete species of the diapente in order and number. **24** At any rate, this was the principal reason that the synemmenon tetrachord, joined naturally through a semitone to the mese step, was devised, **24a** so that [h5v] the diatessaron-melody is produced from the four steps which before the addition of this tetrachord contained the tritone.

25 Since necessity urgently required it, this synemmenon tetrachord was taken away in order to form the perfection of the diapason and diapente, which is completed in the arrangement of this tritone. **26** For musicians preferred the constant struggle with that very bad tritone to disturbing the consonance of the diapason or diapente. **27** The tenth step, designated as the trite diezeugmenon, ascends by a smaller semitone from the paramese; **28** its name is translated as "third of the disjunct steps," and it holds the third position from the mese. **29** The eleventh step is the paranete diezeugmenon, which surpasses the tenth step by the height of a whole tone. **30** The twelfth step is the nete diezeugmenon, which surmounts the eleventh and immediately neighboring step itself by a whole tone; **31** and to the mese, which is called "middle," the nete diezeugmenon step completes the consonance of the diapente. **32** The thirteenth step is the trite hyperbolaion, which surmounts the twelfth step by a smaller semitone. **33** The fourteenth step is the paranete hyperbolaion, which stands apart from the trite hyperbolaion by a whole tone. **34** Then, the fifteenth and final step is the nete hyperbolaion, which exceeds the fourteenth by a whole tone **35** and is distant from the twelfth step by the interval of a diatessaron **36** and relates to the mese, which is the "middle" and which, as the eighth step, completes the consonance of the diapason **36a** and also fills the consonance from the first bisdiapason. **37** It remains worthy to be known that between the hypate meson, which is the fifth step, and the nete diezeugmenon, in the twelfth position, a consonance of the diapason exists, **38** which many regard as a certain example of heavenly

24 Boet. *De inst. mus.* 1.22.214.18–19.
27–34 Boet. *De inst. mus.* 1.22.216.9–16 (table) paraphrased.

order and disjunction. **39** Indeed, they attribute the hypate meson to Saturn, the parhypate meson to Jupiter, the lichanos meson to Mars, the mese to the sun, the paramese to Venus, the trite diezeugmenon to Mercury, the paranete diezeugmenon to the moon, and the nete diezeugmenon to the earth. **40** But the order to Marcus Tullius [Cicero] in the sixth [book] of his *Republic* is the reverse, **41** which leads to the natural effect of having the most remote steps sound together, that is, the one step from the lower end and the other from the higher—**42** for which reason the highest course of the heavenly sphere of stars, whose periodical return is more rapid, is moved by a high and agitated pitch. **43** The course of the moon, which is the lowest, is moved by the lowest pitch. **44** The earth, which remains at rest, always adheres to the lowest place, **45** for which reason Tullius [Cicero] himself posits the earth as silence itself, that is to say, it is immobile. **46** After the earth, he attributes the lowest pitch, that nearest to silence, to the moon, **47** so that the moon is proslambanomenos, Mercury hypate hypaton, Venus parhypate hypaton, the sun lichanos hypaton, Mars hypate meson, Jupiter parhypate meson, Saturn lichanos meson. **48** The farthest star-bearing sphere of heaven [i.e., the sphere of the fixed stars] is mese. **49** Now here is the disposition of the fifteen distinct steps through the appropriate intervals, collected below (fig. 5.3.1).

Chapter **Four**
The measurement of the steps by the tones and the consonances derived from the proportions

1 With the aid of arithmetic, geometricians presented the measurement of the fifteen steps shown above to musicians, who though they crowd tones close together arrange them agreeably and by a fixed law, with arithmetic as a guide. **2** These geometricians rather than the musicians discovered the dynamics of these tones and their measurement. **3** For this reason, when we have derived a measurement divided by the sounding ratios of the string, **4** the greater length of string and the greater numerical quantity produce lower

39 Boet. *De inst. mus.* 1.27.219.7–11: "Namque hypate meson Saturno est attributa, parhypate vero Ioviali circulo consimilis est. Lichanon meson Marti tradidere. Sol mesen obtinuit. Triten synemmenon Venus habet, paraneten synemmenon Mercurius regit. Nete autem lunaris circuli tenet exemplum." Nicom. *Man.* 3.4–8. On the reversal of Mercury and Venus in Nicom., see Roger Bragard, "L'harmonie des sphères selon Boèce," *Speculum* 4 (1921): 206–213. Boet. correlates the order of the planets with the synemmenon tetrachord, whereas Gaf. assigns these same planets to the diezeugmenon tetrachord.
40–42 Cic. *Rep.* 6.17.
43–44 Cic. *Rep.* 6.18; Boet. *De inst. mus.* 1.27.219.17–19.
45–48 Boet. *De inst. mus.* 1.27.219.21–24; Cens. *DN* 13; Cic. *Rep.* 6.18; Pliny *HN* 2.22.84.

pitches; **5** the more contracted length, in contrast, generates the higher pitches by a smaller numerical quantity. **6** By however much a string is either longer or numerically greater or is shorter and denoted by a smaller number, by so much will either a lower or a higher sound be produced. **7** Thus it is necessary to follow physics [*rerum naturam*] in the measurement of the steps and intervals; **8** it follows that the greater numbers are attached to the greater length of strings, from which lowness of pitch results; **9** and the smaller numbers are attached to the shorter length of strings, from which the height of pitch arises. **10** Accordingly, all tones could be produced from a single string, which can be shown in the arrangement above, **11** and every instrument would contain every tone unless one or another instrument were prevented by frets or small bridges from being struck [at any point]. **12** Let us consider these fifteen tones, arranged in a measurable proportion, on a single string in the following manner.

13 Let there be established a string of nine segments; **14** where the first segment begins, one may observe at the bottom the father, as it were, the proslambanomenos, which is the first pitch. **15** Where the first segment comes to an end, we shall establish the second step, that is, the hypate hypaton, which is one tone apart from the first [pitch], **16** because a string of nine divisions contains **16a** only eight segments plus its own eighth part in the ratio of 9:8, **16b** from which a whole tone results. **17** Since the proslambanomenos resides in the first and lower place, it produces the lower tone. **18** From this tone, the remaining tones come forth just as all numbers emerge from unity. **19** Therefore, these two pitches are arranged in this manner, since

10 Cf. Fig. 5.3.1.

11 For a historical overview of the small frets in the music-theoretical literature, see Reckow, "Clavis," *Handwörterbuch* (1971), pp. 1–10.

12 For a graphic representation of Gaf.'s division of the monochord, see Cecil Dale Adkins, "The Theory and Practice of the Monochord," Ph.D. diss., Iowa State University, 1963, p. 210. Further on Gaf.'s use of two sets of Pythagorean string lengths based on the numbers of Aristid. Quint., see Gaf. *Angelicum* 5 and *De harm.* 7.

13 For a survey of various divisions, see Jan W. Herlinger, "Fractional Division of the Whole Tone," *Music Theory Spectrum* 3 (1981):74–83. Gaf.'s nine-part division stands in clear contrast to the five-part division of March. (*Lucid.* 2.5ff.). On March.'s division, see Jan W. Herlinger, "Marchetto's Division of the Whole Tone," *Journal of the American Musicological Society* 34 (1981):193–216; also, Jan Herlinger, "Marchetto's Influence: The Manuscript Evidence," in "*Music Theory and Its Sources: Antiquity and the Middle Ages,* ed. André Barbera, vol. 1 of *Notre Dame Conferences in Medieval Studies* (Notre Dame, Ind.: University of Notre Dame Press, 1990), pp. 241ff.

13–14 Gall. *Rit. can.* 1.6.305.34–39 abbreviated. On the location of notes on the monochord, see also Guido *Microl.* 3–4.

15–16 Gall. *Rit. can.* 1.6.305.1–5 abbreviated.

16a–16b Boet. *De inst. mus.* 5.13.363.23–24.

17–20 Gall. *Rit. can.* 1.6.305.6–9.

no indivisible interval of musical motion occurs between the second and third steps. **20** But it is necessary to derive the fourth step prior to the third, and the interval is less than the half, that is, a smaller semitone; **21** because musical motion that is not complete cannot measure any step with accuracy; **22** hence no larger or smaller semitone can be created by itself. **23** From this results that we divide the whole string into four equal segments. **24** Where the first segment ends, we put the fourth step, that is, the lichanos hypaton, **24a** which answers the first consonance as a diatessaron in the ratio 4:3, **25** for a string of four segments comprises in itself that string which is of three segments and its third part. **26** Now, with the point of the compass placed on this fourth newly created pitch, and having divided the whole string into eight equal segments, let us turn the compass toward the lower part of the string, **27** so that it leads from the ninth segment to the fourth step, **28** where the ninth segment itself ends, and here locate the parhypate hypaton, tuned a whole tone down from the fourth step itself according to the 9:8 proportion. **29** The space [i1r] between the second and third steps derived in this order is fixed; it is the smaller semitone. **30** So these smaller semitones, which are derived either before or after or between the two tones and remain there in a natural order are considered fixed. **31** These are the semitones which are accustomed to arise after the measurement of a tone and from time to time to be dislodged, when they are movable notes. **32** It is well known that every semitone naturally is larger or smaller, **33** since the semitone, whether movable or fixed, in no way divides the string equally in such a manner that the semitone can be derived and produced by means of certain courses of complete musical motions.

34 Again, let us put the point of the compass on the first pitch and partition the whole string into three equal segments. **35** Where the first segment ends, we shall locate the fifth step, that is, the hypate meson, which makes a consonance of the diapente with the first step following the 3:2 proportion, **36** for a string of three segments includes in itself one of two segments and also its own half. **37** With the fourth pitch, this step produces a whole tone. **38** From this point, you should put the point of the compass on the third pitch and divide the remainder of the string into four equal segments. **39** Where you terminated the first segment, you will place the sixth pitch, that is, the parhypate meson, **40** which will produce the consonance of the diatessaron with the third pitch **40a** in the ratio 4:3 **41** and will produce the smaller semitone with the fifth pitch, its neighbor—**42** which could have been produced in no way at all except in this or a similar manner. **43** If, however, the

23–24, 34–42 Gall. *Rit. can.* 1.6.305.10–21 expanded.
43–48 Gall. *Rit. can.* 1.6.305.22–29.

point of the compass is put on the fourth pitch and it marks off through the entire remainder of the string four uniform segments, **44** then where the first segment ends it is necessary to establish the seventh pitch as the lichanos meson, **45** which is distant from the fourth pitch itself in the ratio 4:3 by the interval of the diatessaron, **46** and from the sixth pitch, its neighbor, by a [whole] tone, derived in the ratio 9:8. **47** When we return to the first pitch and divide the whole string into two equal segments, the first segment will lead to the eighth pitch, that is, the mese, which is marked by the proportion 2:1, **48** and completes the consonance of the diapason with the first segment from that place. **49** For a string made up of two segments embraces in itself unity twice. **50** Ptolemy calls this an interval that sounds the same [*aequisone*]. **51** And a bisdiapason arises from this doubled diapason, as has been touched upon before in the second [book], **52** since now the very remote pitches that have been simultaneously set in motion, somehow yield a simple tone from two tones. **53** This is easily perceived by the ears. **54** In this measurement of the string, quantity alone—whether it is longer or shorter—accounts for the difference, **55** since if we choose the longer portion of the string, the shorter portion is doubled. **56** If we choose the shorter portion of the string, it is comprehended twice in the longer; **57** thus the very remote pitches of the diapason also differ only insofar as the low portion is perceived to produce a lower tone and the acute portion a higher tone. **58** For it is the property of equality to be always equal, although, divided up, it changes location. **59** If two parts of a divided continuum are apart in their position, they are nevertheless always equal if their quantity is unchanged. **60** Thus the very remote pitches of the diapason are equal and similar, since through all mingling of the eight tones [i1v] the most remote tones remain paired in likeness. **61** Ptolemy therefore justly calls the diapason an aequisonance [*aequisonantia*], **62** whereas the remaining tones that are alienated from this equality he calls consonances. **63** But the diapente and the diatessaron do not derive from a string divisible by the true middle, **64** and these two intervals are not conceived by the duple ratio; **65** although joined simultaneously, they come together in the union of this equality. **66** Therefore, the diapason is an equality which easily and openly unites and reveals the pitches by a proper and very just measurement in this contractible little string. **67** Nevertheless, the diapason is discerned by the appropriate whole tones and semitones because it may be formed in various ways. **68** Nevertheless, the preferable division of the string at first encloses the eight tones proportionately; **69** from there, the division always progresses to

50 Ptol. *Harm.* 1.7.15.
51 Gaf. *Theorica* 2.2.132.
61 Boet. *De inst. mus.* 5.11.361.30–31; Ptol. *Harm.* 1.7.

the remaining pitches through the diapason. **70** Thus the point of the compass placed on the second pitch divides the string into two equal segments, **71** and where the compass terminates the first segment, place the ninth step, that is, the paramese, which is distant from the eighth by a whole tone, **72** and from this same second pitch the second species of diapason is also derived. **73** Now you do the same thing and affix the compass on the third pitch, **74** where the first segment ends you put the tenth step, that is, the trite diezeugmenon, **75** which is distant from the ninth by a semitone **76** and this produces the third species of diapason. **77** But with the compass placed on the fourth pitch, the sixth step now divides the whole string into two exactly equal segments. **78** The end of the first segment leads to the eleventh step, the paranete diezeugmenon, **79** which is distant from the tenth step by a whole tone and completes the fourth species of diapason from this fourth pitch. **80** When you have done the same thing on the fifth pitch, you will realize the twelfth step, that is, the nete diezeugmenon, **81** which is distant from the eleventh step by a whole tone and produces the fifth species of diapason from the fifth step. **82** Again, after the point of the compass has been placed in the sixth pitch, dividing the string into two equal segments, **83** the first segment, having been determined, adduces the sixth species of diapason on the thirteenth step, that is, the trite hyperbolaion, which is distant from the twelfth step by a semitone. **84** When the point of the compass is put on the seventh pitch and divides the rest of the string into two equal segments, **85** then where the first segment ends the fourteenth step will be obtained, that is, the paranete hyperbolaion, **86** distant from the thirteenth step by a whole tone, which, from this seventh pitch will derive the seventh species of diapason. **87** Finally, the compass is placed on the eighth pitch, that is, the mese. **88** The first segment, which divides the rest of the string into two exactly equal segments, comprises the fifteenth step, that is, the nete hyperbolaion, **89** which is separated from the fourteenth step by a whole tone, and which from the eighth note completes the first replicated species of diapason. **90** The above figure [fig. 5.3.1] includes these divisions of the fifteen steps very clearly, and shows that this note completes the bisdiapason. **91** Consequently, as many tones or steps we may try to derive additionally in this order, we shall derive a step, which can always be divided by a correct half, by this easy procedure.

92 Now if we wanted to derive an additional pitch that occupies a whole tone at the lower end below the first pitch [i2r], it will be possible to divide the whole string into eight equal segments; **93** and after this has been done, turn

70–72 Gall. *Rit. can.* 1.6.305.32–35 expanded.
73–81 Gall. *Rit. can.* 1.6.305.36–39, 306.1–9.
82–86 Gall. *Rit. can.* 1.6.306.10–18 expanded.
87–89 Gall. *Rit. can.* 1.6.306.19–24.

Figure 5.4.1

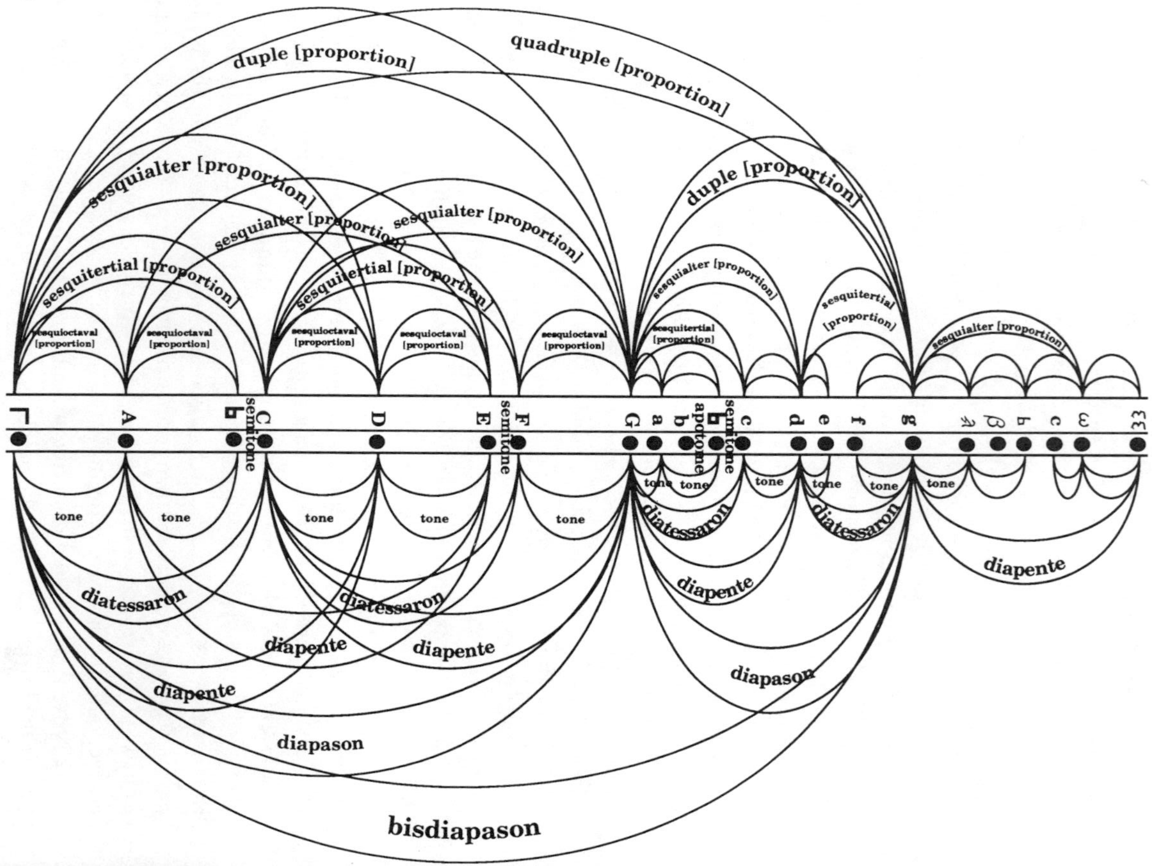

duple [proportion]
quadruple [proportion]
sesquialter [proportion]
sesquialter [proportion]
sesquialter [proportion]
sesquitertial [proportion]
sesquitertial [proportion]
duple [proportion]
sesquialter [proportion]
sesquitertial [proportion]
sesquitertial [proportion]
sesquialter [proportion]
sesquioctaval [proportion]
sesquioctaval [proportion]
sesquioctaval [proportion]
sesquioctaval [proportion]
sesquioctaval [proportion]
Γ
A
♮
semitone
C
D
E
semitone
F
G
a
b
apotome
♮
semitone
c
d
e
f
g
c
ω
tone
tone
tone
tone
tone
tone
tone
tone
tone
tone
tone
diatessaron
diatessaron
diatessaron
diatessaron
diapente
diapente
diapente
diapente
diapente
diapason
diapason
bisdiapason

the point of the compass toward the lower end and mark off one segment beyond the first note by a sesquioctaval proportion. **94** In measuring this ninth segment, you will establish the tone or step and thereby append what the theorists denote by the Greek letter gamma in their musical *Introductorium,* as is made clear by this poem of Guido.

> **95** Certain people put a Greek letter G before the first letter, A, from which the whole string is divided into nine segments. **96** Where the first segment ends will be the first letter.

97 Likewise, following this instruction, everyone will be able to add the remaining letters.

98 If I may summarize in a few words what I said above **98a** the diapason is marked off up to the end always in two segments, the diapente in three, the diatessaron in four, and the whole tone in nine. **99** The greater the number of segments, the smaller will be the interval. **100** Indeed, we cannot traverse the monochord in any other way. **101** The whole tone, which receives its name from the word "tuning," that is to say, "sounding," **102** consists of nine segments as the larger term and eight segments as the smaller term. **103** But the semitone, ditone, and semiditone, although they join pitches, nevertheless cannot divide the monochord by any dimension. **104** Pietro D'Abano, the Conciliator, camer to this same conclusion himself at the end of his eighty-third *Differentia,* **105** for he derives only five consonant proportions from the division of the monochord. **106** To make this clearer, let us here display Guido's *Introductorium,* printed with Latin letters only, and marked with the appropriate tones and proportions (cf. fig. 5.4.1).

Chapter **Five**
The distribution of tones by appropriate semitones

1 Although we have clearly demonstrated the layout of the musical *Introductorium* by means of its consonant intervals, it seems appropriate to show that the individual tones are also divided into two unequal semitones. **2** Since the larger sounding superparticular proportions cannot be equated with the smaller superparticular proportions—for the consonance of the diapente is unequal to the consonance of the diatessaron, and the diatessaron is unequal to the tone—**3** it is likewise necessary to demonstrate that the larger semitones are brought disproportionately together with the smaller semitones.

95–96 Guido *Reg.* (cf. Gerbert, *Scriptores* 2:26, lines 1–3).
98a Guido *Reg.* (cf. Gerbert, *Scriptores* 2:27, lines 19–20).
103a–104 D'Abano *Concil.* 83.
1–72 For a more thorough discussion of the mixed genus, see Gaf. *De harm.* 1.15–16.

4 Assuming the above collection of fifteen neighboring steps from above, let us place the point of the compass in the third step of this arranged and measured string and let us measure off along this whole string eight equal segments. **5** If we turn the compass around at the ninth segment toward the lower part, the end of the ninth segment divides the first tone between the proslambanomenos and the hypate hypaton into a larger and smaller semitone. **6** Because this little note is placed at one's discretion and is removed—**7** due to the fact that it does not occur in the natural and diatonic series of the consonances—no name suitable to it is assigned **8** Therefore, we call it a small bridge [*clavicula*] and name the remaining small bridges with similar names. **9** Accordingly, the first tone, which is divided into two unequal parts, **9a** in the smallest part encloses the smaller semitone **10** and in the larger part the apotome, which, as the larger semitone, is dissonant in itself and unsuitable to simultaneous sounding consonances, **11** though the smaller semitone is the ornament of harmony and has a pleasant temperament; **12** indeed, both connected always fill the whole tone. **13** For truly, that ninth segment, having been formed from the immovable and natural smaller semitone which occurs between the second and third steps **14** and from the larger part, that is, the larger semitone of the first tone divided, yields the tone. **15** It is undeniable that the fixed, immovable, and natural semitone is the smaller; **16** when the larger part of the tone is united with this smaller semitone, the tone is shown to be completed. **17** Here, two semitones have been presented, that is, the tone is divided into a larger and a smaller semitone, as suited to the sesquioctaval proportion. **18** And the smaller semitones are those which are joined to the larger semitones, so that the tone results. **19** Consequently, in the instrument of the monochord, there are several small bridges of this kind, **20** which some call *fictae,* **21** while others think that one must call them *b molles*; **22** still others more appropriately and conveniently call them semitones, **23** since they separate the tones into two unequal parts by their own operation. **24** If you do not bring these unequal parts together by a proper distribution, **25** you conclude that the monochord has been improperly and irrationally derived and divided.

26 Which semitones ought to precede in such a distribution of the tones and

20 For example, Pros. *Par. tract.* 2ff.

21 For example, Con. *Nov. mus.* (De formatione monochordi); for further examples, see Reckow, "Clavis," pp. 1–10.

22 For example, Anon. XI *Tract. de mus. plan.* (cf. Coussemaker, *Scriptorum* 3:419); for an English translation and commentary, see Richard Joseph Wingell, ed., "Anonymous XI (CSIII): An Edition, Translation, and Commentary," 3 vols., Ph.D. diss., University of Southern California, 1973.

whether larger or smaller semitones ought to follow must now be considered. **27** Many consent in the decision of dividing [the tone] by saying that the tone is completed by a smaller and a larger semitone in this manner, **28** just as it is shown that [the tone] is made up by joining a larger and smaller semitone. **29** In such a partition, however, the smaller semitone will be placed more suitably before the larger, that is, situated in the lower part of the tone. **30** This is proved by at least three reasons: **31** first, the argument regarding the tritone. **32** If the whole tone from the mese step is cut down in order to temper the tritone with trite synemmenon and the larger semitone or apotome follows, a new error worse than the previous one arises **33** because the larger semitone followed by two tones is more dissonant than a melody made of three successive tones. **34** Second, if the smaller semitone did not precede the larger semitone, **35** the synemmenon tetrachord either would have been added in vain or would have begun from the worst discord, which under no circumstances could be tolerated. **36** Indeed, it seemed proper that the synemmenon tetrachord itself proceeded by means of a smaller semitone and a tone and a tone, by a diatonic progression just as other tetrachords. **37** Finally, if in such a partition of the tone, the smaller semitone were put after the larger semitone, **38** where the smaller fixed and stable semitone naturally is situated, two continuous smaller semitones would succeed each other: **39** that is, the first being the movable one separated from the preceding tone, **39a** and the second semitone being the stable one. **40** Because these two semitones, joined simultaneously, do not complete the tone, they would lead to rather lascivious and very difficult melody-making; **41** as for example, if you were to place the sixth on the second step of the measured monochord in dividing the whole string into nine equal segments, **42** the first segment would partition the tone between the third and fourth steps into a larger and smaller semitone; **43** and in this manner, as long as you proceeded further through the individual tones, the apotome would always appear to you first. **44** Therefore, you could not escape that inconvenience of bringing together two immediately successive smaller semitones, one after another, **45** which is permitted to occur in certain positions when there is no way out of it, **46** especially when we try to divide this tone which follows the fixed semitone successively into a smaller and larger semitone; **47** for there two smaller semitones succeed immediately. **48** And it is not unseemly to pursue what cannot be avoided.

49 Therefore, two immediately successive smaller semitones are confirmed to succeed each other alternately in such a division of tones. **50** Only, let the smaller semitone precede this natural and fixed semitone and the movable semitone follow. **51** Wherever and whenever the tone is divided to smoothen the unevenness of the tritone, however, under no circumstances

Figure 5.5.1

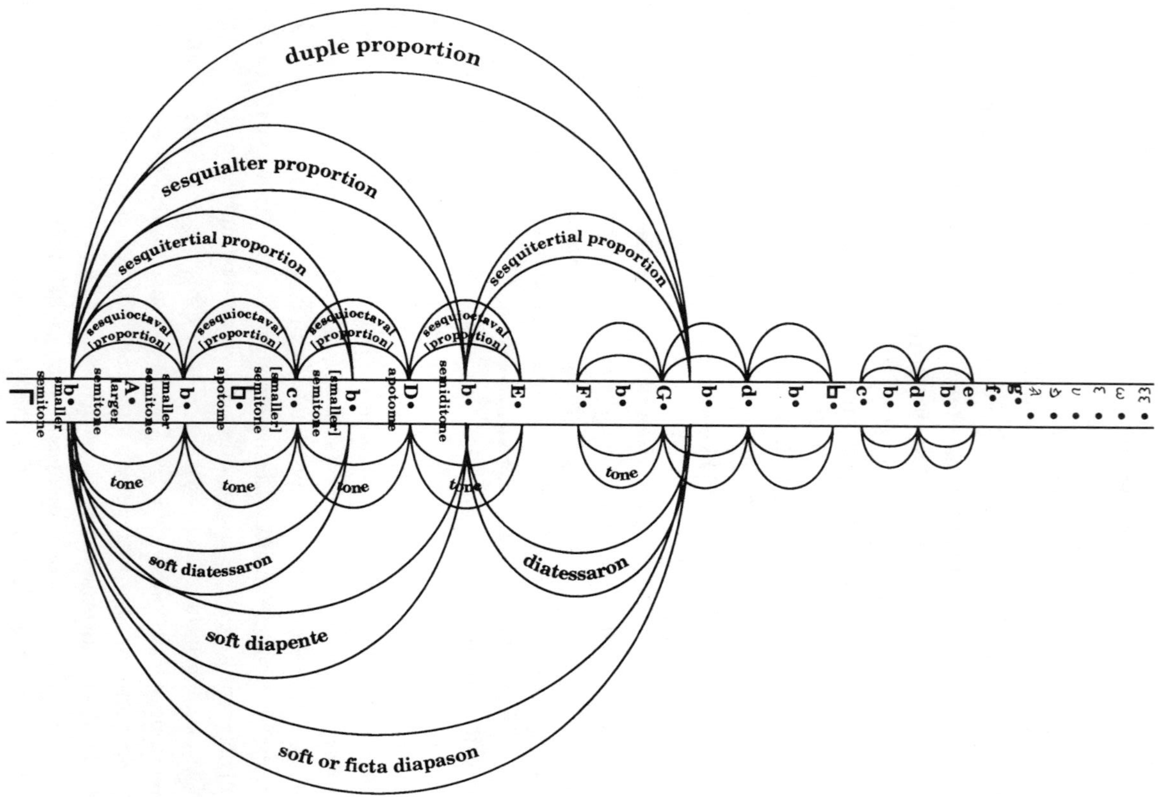
duple proportion
sesquialter proportion
sesquitertial proportion
sesquitertial proportion
sesquioctaval [proportion]
sesquioctaval [proportion]
sesquioctaval [proportion]
sesquioctaval [proportion]
smaller semitone
larger semitone
smaller semitone
apotome
[smaller] semitone
[smaller] semitone
apotome
semiditone
tone
tone
tone
tone
tone
soft diatessaron
diatessaron
soft diapente
soft or ficta diapason

should the larger semitone precede the smaller. **52** In keeping with these assumptions, let the sixth step be placed on this movable little note which divides the first tone in partitioning the whole string into four equal segments; **53** where then the first segment reaches, that is, between the fourth and the fifth step through the diatessaron (through the sesquitertial proportion), **54** there let the small bridge which divides that third tone into a smaller and larger semitones be returned to its former place. **55** Now, on this small bridge just moved dividing the third tone you put your compass [i4r] and traverse the entire rest of the string in four equal segments, **56** and where the first stops, you will place the small bridge producing the diatessaron to that other small bridge with this segment, guided by the sesquitertial proportion. **57** This small bridge indeed shows that the fifth tone between the seventh and eighth steps has been divided into a smaller and a larger semitone. **58** The point of the compass is put on this little note to divide the fifth step and partition the entire rest of the string into two equal segments, **59** and from this spot in which the little note begins, the compass is turned back towards the lower part; **60** where it comes to rest, the small bridge will be placed, dividing the second tone between the third and fourth steps into a smaller and larger semitone, **61** making the consonance of the diapente through the sesquialter proportion with that small bridge from which the compass departed. **62** If we now put the point of the compass on this small bridge which divides the second tone and we partition the whole string into four equal segments, **63** so that where the first segment ends, that is, between the sixth and seventh steps, you place the small bridge, **64** and this divides the fourth tone into a smaller and larger semitone. **65** To that step from which this motion began, there is the interval of the diatessaron, through the sesquitertial proportion. **66** At last, after the point of the compass has been affixed at this small bridge that divides the first tone, traverse the whole string in two equal segments, **67** so that where the first segment ends, that is, between the eighth and ninth steps, it is fitting to place the small bridge **68** which will divide that sixth tone into a smaller and larger semitone, responding to the first small bridge from which you began through the consonance of the diapason. **69** If you likewise divide the whole string into two exactly equal segments from one small bridge to the other, partitioning the tone into two [parts], **70** you will have cumulated the remaining tones through smaller and larger semitones in a clear division (under the guide of the duple proportion and the resulting consonance of the diapason). **71** Since you want to start this partition with that tone which is the step that was added by the Latins [i.e., Γ] in the lower part of the monochord, diligent attention will reveal it. **72** The present figure, which has been engraved with Latin letters, clearly demonstrates the scale of this division (cf. fig. 5.5.1).

Chapter **Six**
The application of letters and syllables to steps according to the Latins

1 The foresight of the Latin musical authors, which imitated the care of the Greek musicians, first brought into use the division of the musical *Introductorium,* marked out by letters. **2** It described the division by means of only the first seven capital letters of the alphabet because of the excellence of their lowness and magnitude. **3** These same letters are replicated in the order of the fifteen arranged steps but made distinct—**4** to differentiate height from lowness of pitch—by a lower case letter, representing the tetrachord of all the remaining highest steps with twin letters in this manner: *A* ♭ *C D E F G a* ♭ *c d e f g* ꜳ **4a** ♮ ɛ ꝺ

5 The first letter, that is, the majuscule *A,* marks the proslambanomenos step, **6** the second letter, the majuscule *B,* serves the hypate hypaton, **7** the third letter, as is easy to see, *C,* furnishes the third step, **8** the fourth letter the fourth step, and so on. **9** Nor will it seem in vain that this tetrachord is marked out by various letters, since we see that many singers reach the same goal of pitch in modulating the voice. **10** They called the first seven majuscules the *graves,* the remaining seven lower case letters, which replicate the majuscules, *acutae,* and the remaining four represented yet again but with certain doubled ciphers as the *superacutae,* having been duplicated by certain extensions. **11** It is true that prior to this arrangement of the letters, scribes indicated the steps by certain very difficult ciphers; **12** foresighted musicians persuaded by its ease, are said to have instituted this other well-used manner of notating these letters around the time of Guido of Arezzo, to which he himself testifies in these verses:

> **13** We have proven that it is best to notate only in letters.
> **14** Nothing is easier than these for the purposes of learning song,
> if they are used frequently for only three months.

15 We read that seven letters alone were arranged, although they bring forth different tones, as Guido himself indicates again in these verses:

1 Cf. above 5.2.40.
2–4 Guido *Reg.* (cf. Gerbert, *Scriptores* 2:25, lines 24–26) expanded.
10 Guido *Microl.* (Table) expanded; for a similar classification, see Pros. *Tract. mus.* B. 116ab.
11 Gaf. does not include the Greek notational symbols in the *Theorica.* On the Greek notation, see Boet. *De inst. mus.* 4.16. Gaf.'s description of the Greek notation as "difficult ciphers" reflects the opinion of Vitr. (*De arch.* 5.3): "Harmonics is an obscure and difficult subject to read and write about, particularly for those who do not know Greek letters" (as translated in: Claude V. Palisca, "Introductory Notes on the Historiography of the Greek Modes," p. 221).
13–14 Guido, *Reg.* (cf. Gerbert, *Scriptores* 2:30, lines 12–14).
15–17 See *Theorica* 5.6.24–31.

16 Seven letters are joined to the respective tones,
and the seven are distinct,
so that the sound of any one step is not the same as any other.

17 These steps are replicated by the same rule and in the same progression, as Guido then shows in these verses:

18 For another group of seven follows afterwards.
19 They are not different but replicated by one rule.
20 For the notes and days equally come in sevens.

21 Musicians have added another letter to those disposed in this order, associating in the last tetrachord a similar letter with a similar sound, thus: ee . **22** They also placed underneath a Greek letter, that is, one which seems first to offer itself in Greek, described by this figure: Γ. **23** Between this letter and the first letter they discover a tone when dividing the whole string into nine segments. **24** Similarly, between the first and second letters, that is, *A* and *B,* there is a tone; **25** but between the second and third letters, that is, *B* and *C,* a smaller semitone; **26** between the third and fourth letters, that is, *C* and *D,* a tone; **27** from the fourth to the fifth letters, that is, from *D* to *E,* a tone; **28** from the fifth to the sixth letters, that is, from *E* to *F,* they bring together a semitone; **29** but from the sixth to the seventh, that is, from *F* to *G,* a tone; **30** from the seventh to the eighth, from *G* to *a,* the latter being the first replicated letter, a tone; **31** from the first replicated letter to [i5v] the second similarly replicated letter, that is, from *a* to ♮, there is a tone; **32** from the second to the third replicated letters, that is, from ♮ to *c,* a smaller semitone, **33** and the division of the monochord above shows very clearly that the remaining letters are arranged in the same order.

34 In this distribution the consonance of the diatessaron always traverses four letters. **35** Since it is true that from the first replicated to the second replicated letters a tone occurs that together with the two preceding tones makes up the tritone in four steps, [the musicians] noted the second letter twice; **36** first in [that] position which is distant from the first step by a single smaller semitone, that is, in the order and process by which the synemmenon tetrachord is added to the other tetrachords; **37** the second they located in that position which is distant from the first by a tone, on the step of the paramese.

16 Guido *Reg.* (cf. Gerbert, *Scriptores* 2:28, line 10).
18–20 See *Theorica* 5.6.24–31; Guido *Reg.* (cf. Gerbert, *Scriptores* 2:25, lines 27–29).
21–34 Guido *Microl.* 4 expanded, based on Boet. *De inst. mus.* 1.4.14; see also Klaus Jürgen Sachs, "Tradition und Innovation bei Guido von Arezzo," in *Kontinuität und Transformation der Antike im Mittelalter: Veröffentlichungen der Kongressakten zum Freiburger Symposion des Mediävistenverbandes,* ed. Willi Erzgräber (Sigmaringen: Jan Thorbecke Verlag, 1989) pp. 233–244, especially pp. 242ff.

38 They are known to designate this differently from the notation of the first location because it is performed with a sharper pitch. **39** They wrote down the first b in this manner; while they notate the second b in this square configuration, ♮.

40 Moreover, the following must be considered; **41** if everything that is sung distinctly can be written down, whatever is written down easily can be adapted to melody; **42** thus it happens that when we read something delightful—since music delights by nature, as the author has made clear more generally in the first [chapter] of the first [book]—once a melody has been applied to it, both those singing and those listening receive greater and more pleasurable delectation from it.

43 Since in grammar every diction is composed of a syllable or, in the vernacular tongue of just one vowel alone, whatever is said and written down, it may be shown, can be sung. **44** If we arrange all the letters of the monochord (omitting the extreme letters, that is to say, the Greek one and the one which is put in the highest spot), and place underneath all the vowels so that every pitch-letter is assigned the same five vowels, we have the following order. (cf. fig. 5.6.1).

45 Now twenty steps, arranged in this manner, **45a** including the two very remote steps are added last, namely, this Greek letter Γ which the Greeks call gamma **46** and we consider equivalent to g, and the one which they placed above the highest step written as follows: ee. **47** The famous monk Guido of Arezzo was led by a certain divine inspiration, examining carefully note by note the hymn of the most holy precursor John the Baptist; **48** he realized that the first syllables from each line were distant from each other by the same intervals, as were the steps as disposed on the monochord of the philosophers. **49** Then, on more careful examination, Guido perceived that a tone existed between the first syllable of the first verse, which is *ut,* and the first syllable of the second, which is *re,* and likewise between this and the first syllable of the third, which is *mi.* **50** Then the first syllable of the third verse to the first syllable of the fourth, that is, *fa,* spanned only by a smaller semitone, **51** while the first syllable of the fifth verse exceeds that of the fourth [16r] by a tone, that is, *sol;* **52** finally, Guido observed that the first syllable of the sixth meter, that is, *la,* departs from the first syllable of the fifth verse by a whole tone. **53** He gathered together in these six syllables the three species of diatessaron, which musicians use frequently in a collection of four single tones arranged in a wonderful and natural and almost divine disposition. **54** Although seven discrete tones are assigned to these seven letters, he thought that only six syllables should be applied to the sounding steps. **55** This modulated division

42 Cf. above, for example, 1.1.58.

55 Guido *Microl.* 17; cf. also Joseph Smits van Waesberghe, "Guido of Arezzo and Musical Improvisation," *Musica Disciplina* 5 (1951):55–63.

Figure 5.6.1

A	B	C	D	E	F	G	a	b	c	d	e	f	g	[illegible]	[illegible]	[illegible]	[illegible]
a	e	i	o	u	a	e	i	o	u	a	e	i	o	u	a	e	i
o	u	a	e	i	o	u	a	e	i	o	u	a	e	i	o	u	a
e	i	o	u	a	e	i	o	u	a	e	i	o	u	a	e	i	o
u	a	e	i	o	u	a	e	i	o	u	a	e	i	o	u	a	e
i	o	u	a	e	i	o	u	a	e	i	o	u	a	e	i	o	u

of high and low syllables musicians distributed among the letters with all vowels, by which means whatever can be spoken is capable of being adapted to melody. **56** Six syllables are known to be written but, naturally enough, varied in this series: *ut re mi fa sol la*. **57** Let the melody and harmony of this hymn be printed as it was notated the first time by means of these syllables:

Figure 5.6.2

Fig. 5.6.1 Guido *Microl.* 17.
Fig. 5.6.2 Guido *Epist.* (cf. Gerbert, *Scriptores* 2:45).

[i6v] **58** It will surely be observed that a semitone is recognized between *mi* and *fa,* and these agree with a certain sweetness, so that these letters in their pronunciation conform to the smaller interval, which does not happen in others. **59** It occurred to Guido that this *Introductorium* or monochord (which many call the hand because it is frequently perceived in the left hand) ought to be fitted with these six syllables in their wonderful arrangement. **60** Just as in the distribution of the letters—that is, between *b* and *c, E* and *F,* and sometimes *a* and *b,* the harshness or roughness of the tritone appears—the smaller semitone is perceived to occur in a diatonic progression; **61** thus, with this wonderful series of syllables and its unique pronunciation and disposition (something that is quite elegant and excellent), Guido has put all the smaller semitones in their place. **62** Hence, he applied the set of these six syllables in a sevenfold manner, inserting a tone between the first and second syllables, that is, *ut re,* a tone between the second and third, *re mi,* a tone between the third and fourth, *mi fa,* a smaller semitone, however, between the fourth and fifth, *fa sol,* and a tone between the fifth and sixth, *sol la.* **63** Therefore, Guido assigned these syllables to the letters in this order: **64** to the first Greek letter, [Γ], he applied the first syllable; Γ *ut* [*gamma ut*]; **65** to the first Latin letter, *A,* which is the second step, he assigned the second syllable, *A re;* **66** to the third step, which is described by the second Latin letter [*B*], he attached the third syllable; *B mi;* **67** to the fourth step of the monochord, to the letter *C,* he tacked on the fourth syllable; making *C fa.* **68** Then, to this letter and step, he joined the first syllable, *C fa ut.* **69** To the fifth step, which applies the fourth Latin letter, he joined the fifth syllable, *sol,* and also the second syllable of the second *deductio,* which began on the preceding neighboring step [*C*], that is, *D sol re.* **70** On the sixth step occupied by the fifth Latin letter, Guido established the sixth syllable, that is, *la,* and the third syllable of the second deductio, making *E la mi.* **71** With the sixth letter, which occupies the seventh step, he placed the fourth syllable of the second deductio and the first syllable of the third deductio, namely, *F fa ut.* **72** For the seventh letter, which holds down the eighth step, he chose the fifth syllable of the second deductio, the second syllable of the third deductio, and the first syllable of the fourth deductio in this way: *G sol re ut.* **73** Thus the syllable *ut* is the head and the beginning of all subsequent syllables, that is, of *re mi fa sol la,* in whatever deductio and the set of those six syllables is called a deductio in the harmonic system, for it is perceived as a drawing out of pitches from the low to the high and from the high to the low. **74** Every deductio unites three diatonic species of diatessaron.

75 With the first replicated letter, which leads the ninth step, Guido connected the sixth syllable of the second deductio, the third syllable of the third

59 Cf. above 5.2.40.

deductio, and the second syllable of the fourth deductio in this way: *a la mi re*. **76** In the second repeated letter, which falls on the second step, the third syllable of the fourth deductio is assigned between this letter and the preceding one, by a letter with the same name—but indeed deviating from each other in letter and property, since the fourth syllable of the third deductio has been disposed in this usage, namely, *b fa* [k1r] ♮ *mi*. **77** In the third replicated letter he combined the fifth syllable of the third deductio, the fourth syllable of the fourth deductio and also the first syllable of the fifth deductio, in this manner: *c sol fa ut*. **78** To the fourth repeated letter, he assigned the sixth syllable of the third deductio, the fifth syllable of the fourth deductio, and the second syllable of the fifth deductio as here: *d la sol re*. **79** In the fifth repeated letter, which brings on the thirteenth step, he placed the sixth and last syllable of the fourth deductio and the third syllable of the fifth deductio in this way: *e la mi*. **80** In the sixth repeated letter, which marks the fourteenth step, he arranged the fourth syllable of the fifth deductio, and the first syllable of the sixth deductio in this way: *f fa ut*. **81** To the seventh replicated letter, which corresponds to the fifteenth step, Guido attached the fifth syllable of the fifth deductio, the second syllable of the sixth deductio, and the first syllable of the seventh deductio, adding up to *g sol re ut*.

82 Again, to the first triplicated letter, which corresponds to the sixteenth step, he placed the sixth syllable of the fifth deductio, the third syllable of the sixth deductio, and the second syllable of the seventh deductio in this order: 𝔞 *la mi re*. **83** To the second triplicated letter, which corresponds to the seventeenth step, he placed the third syllable of the seventh deductio in this manner: ♮ *mi*. **84** And also, between this letter and the preceding one, Guido arranged another second letter, different in terms of property and configuration, in order to divide this tone with the fourth syllable of the sixth deductio in this manner: *B fa* ♮ *mi*. **85** Moreover, to the third triplicated letter, which corresponds to the eighteenth step, he placed the fifth syllable of the sixth deductio and the fourth syllable of the seventh deductio, as shown here: 𝔠 *sol fa*. **86** To the fourth triplicated letter, which corresponds to the nineteenth step, he placed the last syllable of the sixth deductio in this manner: ω *la sol*. **87** Finally, in bringing the twentieth step in contact with the fifth triplicated letter, he placed the last syllable of the seventh deductio in this manner: 𝔢𝔢 *la*.

88 From this *Introductorium* musicians were accustomed to notate the measured disposition of the syllables and letters, and some consonances by certain square notes, on lines and in spaces, as the present figure [fig. 5.6.3] demonstrates.

89 We thought that this diagram of letters and syllables measured out through proper tones and fixed semitones should be published as a musical *Introductorium,* since it is introduced at the outset to novices, just as the

Figure 5.6.3

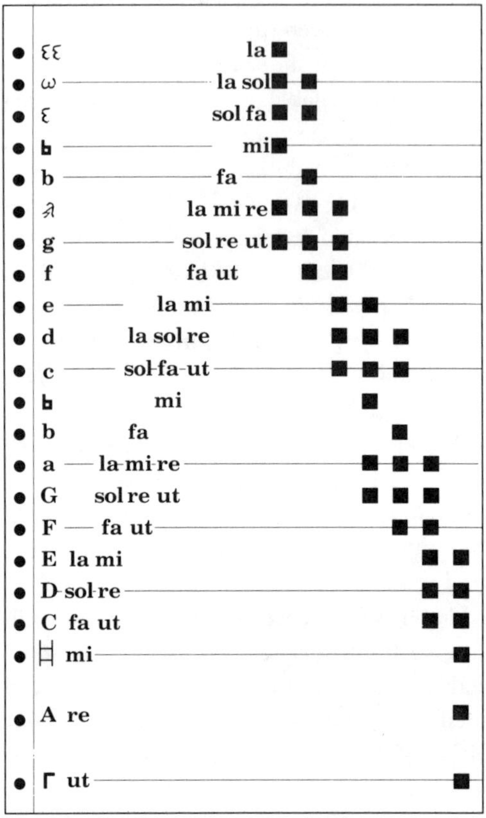

grammarians are accustomed to teach the alphabet to boys. **90** If anyone were to try to place other syllables or signs and letters in the lower part, **90a** he would join *G sol re ut* and below this *F fa ut* **90b** and then *E la mi,* and in this manner the remaining letters by replication. **91** Also, if someone wished to place letters above the higher part [i.e., the *acutae*], he would put on ee *la* ee *la mi*; **92** from there, he would proceed higher to *F fa ut* and afterwards to *G sol re ut* and then to *a la mi re*; **92a** and so on, replicating according to the former progression and arrangement. **93** Even if these syllables are used in the introduction to singing, musicians hardly agree to assign and ascribe them to the diagram of consonances, because very often either words confuse syllables or syllables confuse words in an unclear diagram. **94** Therefore the lowness and height of these syllables, arranged alternately on lines and in spaces, are notated by special little square, diamond, and other shapes of notes, **95** but

95 Gaf. *Pract.* 2.2–5.

there will be a fuller account of these notes in the second part, that is, in the *Practica*.

96 Likewise, the [musicians] think that a note is very easily signified by a syllable and vice versa. **97** In addition, there are three properties of these seven *deductiones,* that is, *quadralis,* named after the square shape, and *b rotundum,* after its shape. **98** They are also known by the quality of the tone itself as b *durus* [hard] and *b mollis* [soft]. **99** The third property is natural. **100** Marchetto defines this property as a derivation of many pitches from one and the same principle. **101** Some theorists assert that this property is an essential or accidental acceptance of the pitches seated in the low or high register. **102** Certainly, they write notes for syllables, and the syllables are the names of steps and pitches. **103** Also, pitches cannot be notated but are retained in memory, lest they perish. **104** The lowness and the height of notes are accurately recognized to refer to lines and intervals in the notation of melodies. **105** For that reason the properties are considered in this progression. **106** Every deductio of these six syllables mentioned above that has its beginning and origin in the letter *G* is said to be of the property of ♮ *quadralis* or *duralis.* **107** A deductio that begins on the letter *C* is called *naturalis* [natural]; **108** one that begins on the letter *F* **108a** is called *b mollis* or *b rotundum,* concerning which there is a fuller discussion in the fifth [book] of the *Flos musicae* **109** and a copious consideration in the first book of the *Practica.* **110** Thus, since there are seven deductiones of the syllables, three of them are accustomed to be assigned to the property of ♮ *quadralis,* the first of which begins on the note that follows the initial Greek letter Γ [gamma] (so named because it sounds like the letter g). **111** The remaining two deductiones are in the other two letters g. **112** Two deductiones are assigned to the natural property, **113** and they originate in *C fa ut* and in *c sol fa ut.* **114** Those belonging to the *b mollis* property are those two deductiones which begin on the two *F fa ut.*

Chapter **Seven**
The species of diatessaron and diapente

1 It is well known that the consonance of the diatessaron is composed of four steps, comprising two whole tones and one smaller semitone, as has been

98 Ptol. *Harm.* 1.12, 15–16. On Ptol.'s terminology (syntonon = durum; malakon = mollis) as well as on the use of these terms in the Latin West, see Carl Dahlhaus, "Die Termini Dur und Moll," *Archiv für Musikwissenschaft* 12 (1955):280–296.
100 March. *Lucid.* 8.3.3; see also Gaf. *Pract.* 1.4.
105–108 Gaf. *Pract.* 1.4.
108a Gaf. *Flos mus.* 5 (lost).
109 Gaf. *Pract.* 1.4.
1 Cf. above 4.4.19. The species of diatessaron are also discussed in Gaf. *Pract.* 1.5.

explained more fully, **2** and this consonance produces three different species, because it is filled up completely by only four pitches, for, as we have previously stated, every consonance always has one species fewer than it has pitches.

3 The first species of diatessaron is made up of a whole tone, a semitone, and a whole tone, proceeding by the syllables *re mi fa sol,* **4** and this species of diatessaron is begun on *A re,* which is tacked on to the proslambanomenos step; **5** wherever a similar progression occurs, it is always indicated as the first species of diatessaron. **6** The second species of diatessaron goes up by a semitone, a whole tone, and a whole tone, produced by the syllables *mi fa sol la*; **7** this species of diatessaron is indeed begun on the step ♮ *mi,* evidently the third step from the introductory step. **8** But the progression of similar syllables always produces the second species of diatessaron. **9** The third species of diatessaron is derived by going up a whole tone, a whole tone, and a semitone in these syllables, *ut re mi fa*; **10** for this species of diatessaron begins on the fourth step of the *Introductorium,* that is, on *C fa ut*; **11** and if the species advances by means of similar syllables or pitches, the third species of diatessaron always occurs. **12** Thus this consideration leads to a relaxation contrary to tension, as is obvious to the more conscientious observer.

13 Moreover, it is evident that every single consonance can have at most as many intervals as are comprised in its species, **14** or it contains one interval less than the number of its pitches; **15** and this happens when all steps of whatever consonance are arranged successively. **16** Whenever an interval interjected between two extreme steps is so disposed that the three syllables *re mi sol* or *re fa sol* may be sung, then it will consist of a consonance of the diatessaron.

17 The remaining consonances advance in different intervals, under a similar principle; **18** and the fewer the steps the fewer will be the intervals, but they will be larger, as everyone can easily perceive. **19** Thus the consonance of the diapente is completed by five steps, comprising three whole tones and one smaller semitone. **20** Since it has five steps, as shown, it produces altogether four different species. **21** Of these four species, **21a** the first is made from the first species of diatessaron and one superimposed whole tone **21b** so that it proceeds by the syllables *re mi fa sol la*; **22** the species of diapente begins with the fifth step of the *Introductorium,* that is, on *D sol re,* **23** and this species is terminated with the ninth step, that is, *a la mi re*; **24** a progression anywhere of similar syllables is always attributed to the first species of diapente.

2 *Theorica* 5.2..35
3–12 Cf. Boet. *De inst. mus.* 4.14.
14 Cf. also above 5.2.35.
21 The species of diapente are also discussed in Gaf. *Pract.* 1.6, and *De harm.* 2.31.

25 The second species of diapente is derived from the second species of diatessaron and one whole tone, that is, in the syllables *mi fa sol la mi*. **26** This species of diapente acquires its beginning in the sixth step of the *Introductorium*, that is, *E la mi* **27** and this species is concluded with the tenth step, ♮ *mi*, where the philosophers are accustomed to place the paramese. **28** A union of five similar steps always indicates the second species of diapente.

29 The third species of diapente is arranged from the tritone and the semitone; **30** this species of diapente originates from the seventh step of the *Introductorium, F fa ut,* to the eleventh, *c sol fa ut,* and in the syllables *fa sol re mi fa,* **30a** whose progression always constitutes the third species of diapente.

31 The fourth species of diapente is formed from the third species of diatessaron and one superimposed whole tone, so that it proceeds from the eighth step of the *Introductorium, G sol re ut,* **31a** to the twelfth, *d la sol re,* that is, **31b** in the syllables *ut re mi fa sol*; **32** and wherever a progression of similar syllables arises, the fourth species of diapente will always result.

33 If you tried to arrange the progression of these species of diapente continuously, proceeding from the second step of the *Introductorium* from which the first species of diatessaron begins, you will derive as the subsequent species of diapente an imperfect one, which by nature is arranged of two whole tones and two smaller semitones; **34** and you would ascertain also that the subsequent third species of diatessaron that you would obtain is similar to the fourth species of diapente, and that the fourth species of diapente obtained is not all unlike the first species of diapente. **35** Concerning these [matters], much is dealt with more copiously in the first [book] of the *Practica,* which is devoted to plainchant. **36** Among the different intervals, the consonance of the diapente is completed and derived through various movements, as is established by a more careful investigator.

Chapter **Eight**
The species of the consonance of the diapason

1 The consonance of the diapason is completed by eight steps which contain five whole tones and two smaller semitones, **1a** and is named "di-

35 Gaf. *Pract.* 1.5–6.

1 On the earlier discussion of the species of diapason, see above 5.4.72ff. The species of diapason are also discussed in Gaf. *Pract.* 1.7.

1a On descriptions and etymological derivations of the diapason, see Joseph Smits van Waesberghe, "Die Interpretation von *Diapason* in den mittelalterlichen Traktaten," trans. M. U. Schouten-Glass, in *Dia-pason de omnibus-Ausgewählte Aufsätze von Joseph Smits van Waesberghe: Festgabe zu seinem 75. Geburtstag,* ed. C. J. Maas and M. U. Schouten-Glass (Buren: Frits Knuf, 1976), pp. 41–47.

apason," which means "through all," **1b** because this consonance is completed by all the steps plus one superimposed, **1c** for there are altogether seven tones, **1d** as has been explained more fully in the preceding chapters. **2** Moreover, the diapason is made up of the diapente and the diatessaron, which, since they have many and diverse species, likewise circumscribe the consonance of the diapason in diverse species. **3** Certainly, being formed of eight steps, it comprises only seven different species.

4 Of those, the first species of diapason is made from the first species of diatessaron, which goes from *A re* to *D sol re,* and from the first species of diapente, which is made from *D sol re* to *a la mi re*; **5** and so, the first species of diapason extends from *A re* to *a la mi re,* or from proslambanomenos to mese. **6** The second species of diapason is constructed from the second species of diatessaron which is between low *B mi* and low *E la mi,* and from the second species of diapente, which goes from low *E la mi* to high ♮ mi. **7** Thus the second species of diapason occurs between low *B mi* and high ♮ *mi,* or between hypate hypaton and paramese. **8** The third species of diapason is formed from the third species of diatessaron, which is enclosed between low *C fa ut* and low *F fa ut,* **8a** and from the third species of diapente, proceeding from low *F fa ut* to high *c sol fa ut.* **9** And this third species of diapason proceeds **9a** from *C fa ut* **9b** to *c sol fa ut,* or from parhypate hypaton to trite diezeugmenon. **10** The fourth species of diapason is united from the first species of diapente, which is made from *D sol re* **10a** to *a la mi re,* namely, from lichanos hypaton to mese, and from the first species of diatessaron, which is reckoned from *a la mi re* to *d la sol re.* **11** Thus this fourth species of diapason stretches from *D sol re* to *d la sol re,* or from lichanos hypaton to paranete diezeugmenon. **12** The fifth species of diapason comprises the second species of diapente between low *E la mi* and high ♮ *mi,* or between hypate meson and paramese, **12a** and the second species of diatessaron between high ♮ *mi* and high *e la mi*; **13** and so the fifth species of diapason extends from low *E la mi* to high *e la mi,* or from hypate meson to nete diezeugmenon. **14** The sixth species of diapason is assembled from the third species of diapente, which occurs between low *F fa ut* and *c sol fa ut,* **14a** or between parhypate meson and trite diezeugmenon, and from the third species of diatessaron, which is enclosed between high *c sol fa ut* and *f fa ut,* or between trite diezeugmenon and

1d *Theorica* 5.4.13ff.
3a For a discussion of the seven octave species, see also Gaf. *De harm.* 2.32.
4 Gall. *Rit. can.* 3.5.335.38.
6, 8 Gall. *Rit. can.* 3.5.335.5–16 abbreviated.
10 Gall. *Rit. can.* 3.5.335.18–19 abbreviated.
12 Gall. *Rit. can.* 3.5.335.23–24 abbreviated.
14 Gall. *Rit. can.* 3.5.335.28–30 abbreviated.

trite hyperbolaion. **15** Thus the sixth species of diapason extends from low *F* to high *f*, or from parhypate meson to trite hyperbolaion. **16** The seventh and last species of diapason is coupled from the fourth species of diapente, which is made from low *G sol re ut* to high *d la sol re,* and from the first species of diatessaron, which is comprised between high *d la sol re* and high *g sol re ut*; **17** and the progression of this seventh distinct form of diapason is made from low *G sol re ut* to high *g sol re ut,* or from lichanos meson to paranete hyperbolaion. **18** The consonance of the diapason comprises at most seven intervals when, that is, it is derived from all its eight steps. **19** But it is produced very frequently by six or five and likewise in various ways by four or three or two "diastemata." **20** Finally, the diapason is formed by one interval when it is arranged by the response of only the extreme pitches, that is, of the lower and higher pitches after the complete intervals of five whole tones and two smaller semitones have been interposed.

21 The first [book] of our *Practica* reveals and discusses more fully these four *maneries,* that is, the *protus, deuteros, tritos,* and *tetrardos* and their plagal versions, **21a** which musicians used to maintain in making any melody.

22 These seven species of diapason are placed in this system in such a way that the first three require the consonance of the diatessaron below and the diapente above the diatessaron; **23** the remaining four species are arranged in the contrary manner with the diapente below, as previously described.

24 The philosophers called these seven species of diapason "modes," from "modulans" **24a** or "moderans," as they perceived these [modes] to moderate all progressions of melody through fixed limits of high and low pitch. **25** They called the first species of diapason, from proslambanomenos to mese or from *A re* to *a la mi re,* the Hypodorian. **26** When all pitches of the Hypodorian are raised by a [whole] tone, they produced the second mode, the Hypophrygian. **27** When all the pitches of the Hypophrygian are [raised] by a semitone, they form the Hypolydian. **28** All the pitches of the Hypolydian, brought up by a whole tone, constituted the Dorian. **29** When all the pitches of the Dorian were brought up a whole tone, they constituted the Phrygian. **30** Again, all the pitches of the Phrygian, [raised] by a semitone, produced the Lydian. **31** After all pitches of the Lydian had been tuned up by a whole tone, the Mixolydian was produced. **32** Therefore, these seven modes demanded

16 Gall. *Rit. can.* 3.5.335.33–38 altered.

21 Gaf. *Pract.* 1.8–15.

23 Cf. above 5.8.4ff.

24 This remark signals a shift from the plainchant theorists to Boet., who called the transpositions *modi* (cf. *De inst. mus.* 4.15–16). Gaf.'s scheme of transposition—different from that reported by Boet. (cf. *De inst. mus.* 4.15ff.: tone, tone, semitone, tone, tone, semitone, tone)—can be justified in the context of the Guidonian gamut.

those same steps that the first species of diapason was shown to possess, as from proslambanomenos to mese, or from low *A* to high *a.*

33 One mode surmounts another either by a tone or by a semitone, as the diatonic arrangement of the *Introductorium* has been shown to possess naturally. **34** Thus the proslambanomenos or *A re* of the Hypophrygian exceeds that of the Hypodorian by a whole tone. **35** Similarly, the mese or *a la mi re* of the Hypophrygian exceeds that of the Hypodorian by a whole tone. **36** Then the intervening pitches and also the complete order of the Hypophrygian surpass the remaining interposed pitches and the complete order of the Hypodorian by the interval of a whole tone. **37** Likewise, this order and also this process occur in the others, as everybody can easily comprehend for himself. **38** It is well known that, accordingly, the seven separate modes are derived according to the seven species of diapason in the same steps and pitches, standing apart only by height and depth. **39** To these, Ptolemy himself added at the top the Hypermixolydian, whose progression of all pitches surpasses the disposition of the Mixolydian by a [whole] tone. **40** We thought that the arrangement of these modes so derived ought to be printed here in a diagram (fig. 5.8.1) with the first eight letters of the *Introductorium* in this order.

41 Our Boethius called these modes "tropes," which he concludes to be systems of entire series of pitches differing by height and depth. **42** [Further-

33 Cf. Guido *Microl.* 7.

34–37 Gaf. reports the functional (or dynamic) nomenclature given in Boet. *De inst. mus.* 4.17.

39 Boet. *De inst. mus.* 4.17.348.2–3; Ptol. *Harm.* 2.10.63.5–7. The association of Ptol. with the eighth mode is also discussed in Bryen. *Harm.* 1.8. On this attribution, see Bower, trans., *Boethius: Fundamentals,* p. 160.

41–42 Boet. *De inst. mus.* 4.15.341.21–22; Nicom. lost *opus maior* (cf. Calvin M. Bower, "The Modes of Boethius," *Journal of Musicology* 3 (1984):252. The sentence preceding this remark of Boet. is crucial for the understanding of this passage. "Species autem est quaedam positio propriam habens formam secundum unumquodque genus in uniuscuiusque proportionis consonantiam faciendis terminis constituta" (Boet. *De inst. mus.* 4.14.337.22–25; as translated by Bower in *Boethius: Fundamentals,* p. 153: "From the species of the consonance of the diapason arise what are called 'modes.' They are also called 'tropes' or 'tones.' "). For an explication of this passage, see Palisca, *Humanism,* pp. 39ff. The incorrect interpretation of this sentence (not cited in the *Theorica*) results in Gaf.'s confusion of the terms octave species, modes, and tonoi—terminology which is clarified in the later *De harm.* 2.32. On the use of this terminology in Gaf.'s *Theorica* and *De harm.,* see Palisca, *Humanism,* pp. 293ff. For a more general overview of these concepts during antiquity and the early Middle Ages, see André Barbera, "Octave Species," *Journal of Musicology* 3 (1984): 229–241; Bower, "The Modes of Boethius"; and Jon Solomon, "Towards a History of *Tonoi,*" *Journal of Musicology* 3 (1984):242–251. On the clarification of this terminology in subsequent passages of Boet. *De inst. mus.* 4.15ff. and also on the tenth-century anonymous treatise *Mus. ench.,* see Charles M. Atkinson, "On the Interpretation of *Modi, quos abusive tonos dicimus,*" in *Hermeneutics and Medieval Culture,* ed. Patrick J. Gallacher and Helen Damico (Albany, N.Y.: State University of New York Press, 1989), pp. 147–161.

Figure 5.8.1

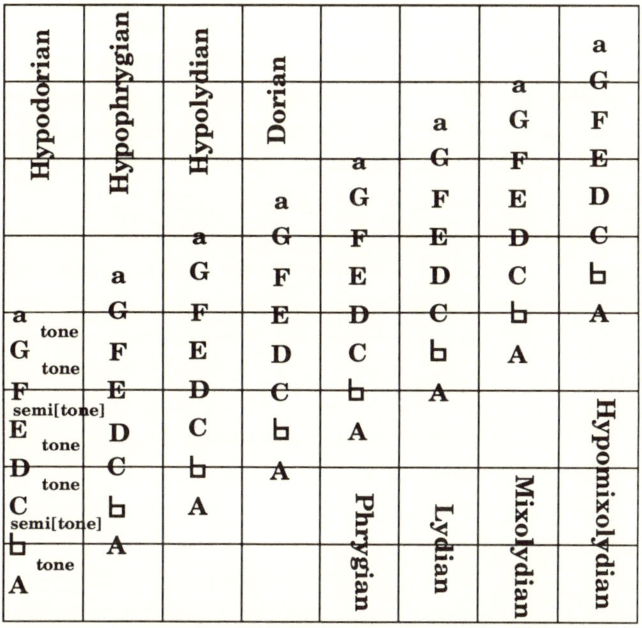

more, Boethius concludes] that a system is the entire corpus of melodic material [*modulatio*], existing through the connection of consonances. **43** "Tropes," therefore, are so called because of the conversion by which one trope is derived from another by turning from depth to height; **44** for "tropos" in Greek is translated as "conversio" [conversion, change] in Latin.

45 On the other hand, Bacchius, a more recent interpreter of the musical science among the Greeks, places only seven modes against the system of the seven species of diapason. **46** Yet in a letter to Boethius, Cassiodorus recalls five, which he calls tones [*tonoi*] and names them according to the place or province where each was found and where the native people were most attracted by their delight in it. **47** Cassiodorus asserts that the Dorian is the

44 For a brief synopsis of modal terminology, see Heinrich Hüschen, "Der Modus-Begriff in der Musiktheorie des Mittelalters und der Renaissance," *Mittellateinisches Jahrbuch* 2 (1965): 224–229.

45 Bac. *Intro.* 77.

46 Cassiod. sent the letter to Boet. in 506, asking him to send a kitharist for the Frankish King Chlodwig. For the content of this letter, see Wille, *Musica Romana,* pp. 700–704. On the characterization of these five tones, see Apul. *Flor.* 3.17 and Pl. *Pol.* 398f.

47–51 Cassiod. *Var.* 2.40.32–38; see also: Cassiod. *Inst. hum.* 2.5. For a fuller discussion of the modal ethos, see Gaf. *De harm.* 4.2–5.

granter of judgment and also the insurer of chastity. **48** The Phrygian, as he says, stimulates fighters and inflames the desire for violence. **49** The Aeolian calms the tempests of the soul, having already granted sleep to those who have been calmed. **50** He claims that the Iastian sharpens the intellect of dull people, especially as it grants an appetite for celestial goods. **51** The Lydian releases the too great cares of the soul and boredom by its renewing quality and strengthens the soul through delight. **52** To these five excellent and principal modes, ten others were added so that every single one of the five would embrace two related modes, as Cassiodorus seems to declare.

53 Martianus [Capella] reckons these modes in another order, saying that the Lydian is the first mode, with which the Hypolydian and the Hyperlydian are joined. **54** The second is the Iastian, whose companions are the Hypoiastian and the Hyperiastian. **55** The third is the Aeolian, with which the Hypoaeolian and the Hyperaeolian unite. **56** The fourth is the Phrygian, with the Hypophrygian and the Hyperphrygian. **57** He asserts that the fifth is the Dorian, with which the Hypodorian and the Hyperdorian unite.

58 These modes or tones [*tonoi*] also produce various affections according to their complexions, disposed by nature out of diverse things. **59** For one mode excites by high pitch, and another calms by low pitch (as has been more fully shown in the first [chapter] of the first [book]). **60** Similarly, one mode imparts joyfulness, another grief; **61** and so it has been learned that individual modes create diverse effects. **62** But they say that the human voice, fit in itself and trained by art, is far more fit than the sounds of all instruments to soothe the ears, to calm the passions of the soul itself, to expel feebleness of the body, and to excite motions and to alternate impetuosity. **63** Certainly, by this means and melody, words, and the meaning of these words are thought to be represented. **64** Therefore, the soul is moved by a twofold affection, that is, by simultaneous harmony and by the flow of the words—with the greatest possible measure, order, and harmonious number—to soothe these human ears with wonderful sweetness, **65** what Aristotle appears to have disputed and concluded very firmly in the musical *Problems*.

66 Every tone [*tonos*] or mode has a highest and a lowest point which are

53–57 Cap. *De nup*. 9.935; Aristid. Quint. *De mus*. 1.4–19 (cf. Palisca, *Humanism*, p. 42).
58 On the tonoi, see Ptol. *Harm*. 2.5. Nicolo Leoniceno, who completed his translation of Ptol.'s *Harm*. (prepared for Gaf.) in 1499, was the first Western author to transmit this theory (cf. Palisca, *Humanism*, p. 43). Also crucial for Gaf.'s clarification of terminology was the publication of Giorgio Valla's translation of Cleon.'s *Intro. Harm*. in 1497—both volumes which were accessible to Gaf. in the preparation of his *De harm.*, though presumably not for the completion of the *Theorica*.
59 Cf. above, for example, 1.1.80.
62–64 D'Abano *Expo. Pr. Arist*. 19.15 paraphrased.
66 D'Abano *Expo. Pr. Arist*. 19.20; possibly also 19.38 (cf. Palisca, *Humanism*, p. 197).

brought together in the midpoint [*mese*]; truly, since without the modes themselves, those things cannot be brought together to themselves by an alternate interchange. **67** It is not an unprofitable invention to have an exquisite artificial music discovered by the investigations of composers on diverse instruments and lyres to be contained in the same tones [*tonoi*] or modes, which human ingenuity in very learned inquiries has perceived to understand, **68** that is, a certain harmony of the diapason, assembled from all, which wonderful union retains all the very celebrated virtues of simultaneous harmony. **69** Indeed, the kitharist ornaments his own kithara proportionately with strings, with these alternate modes, and with this harmony. **70** If he altered this kithara to another mode, or relaxed these strings and made the whole order lower, or tightened them and made it higher (nonetheless retaining always the diapason and its whole order), then the harmony would be led from one mode to another. **71** The performer on his lyre as well as other musicians on their instruments do the same. **72** Hence, I believe that from this arise these various effects of the diverse modes and the various passions of men and of all living beings, for [Aristotle] in his *Poetics* conceded to each one according to his conviction to wish his own, because we do not live by one vote alone. **73** Nor does it seem that the one and the same method of making melody is used by all. **74** Englishmen shout joyfully when making harmony, **74a** the French sing, the Spaniards wail, Germans howl; **75** and they say that some Italians, such as the Genovese and those who reside on that coast, bleat like goats. **76** Yet, the Barbarians attribute barking to other peoples (to which I take exception). **77** Later generations of the Christians nonetheless respect these Italians as most famous in the discipline of the art.

78 Church musicians call these modes, of which there are eight, tones, as if they have been derived and determined from the tones or parts of tones, that is, semitones. **79** They say, indeed, that these [tones] are the rules through which all modulated harmony is recognized, distinguished, and judged by the height and lowness [of pitch] at the end [i.e., the final]. **80** We have set forth the progressions and formulas of these [modes] more fully in the first [book] of the *Practica*.

72 Arist. *Poet.* 1448b.II.20–21.
74a The origin of this citation cannot be determined. For later sources to include this remark, see Aaron *Lucid. in mus.* 4.1 (kindly identified by Dr. Bonnie J. Blackburn) and Finck *Pract. mus.* 5.71–72 (kindly identified by Peter Slemon).
79 On the significance of range and finalis in the classification of the modes, as related by plainchant theorists, see Dolores Pesce, *The Affinities and Medieval Transposition,* part of *Music: Scholarship and Performance,* ed. Thomas Birkley (Bloomington and Indianapolis: Indiana University Press, 1987).
80 Gaf. *Pract.* 1.8–15.

THE POEM OF LANCINUS CURTIUS

Let us proclaim in verses the divine descent of Terpsichore,
who improves the melodies of the people of Pieria,
the reward of invention. I do not now prepare to lift
the ivory of the gilded tortoise-shell lyre to life
or the auloi of the raving Satyrs in song,
nor to speak the names of magpies to the woods,
but the name of her who applied honeyed melody to sweet sounding kitharas, yes,
in order that a thin line confine
the space, the notes, and their usage.
This established norm keeps the harmonious from the dissonant ones.
And a firmer rule moderates
what sets the sounding bodies in motion by vibration.
Or when the rule is violated, what can be sung,
detaining those participating. The companions whom she draws
by outstanding light of the stars, are equals. And from the beginning
she traverses with a pair of compasses the measure of infinite things.
Subsequently she counts what is almost uncountable [k5v].
Through the darkness she looks upward to the heavens.
To those who wish to try the paths to the heavens,
she has shown the light and the way to this land of Elysium
and to the home of the gods. She places the origin of Elysium
and its neighborhood in the pole of the celestial vault;
where the image of the lyre glitters together as bright lode stars.
She tempers the voices of orators with sound
just as a slave did for Gracchus; she is suitable to arts.
Or she is a pleasant companion; she composes divine poems,
she praises Olympic rewards:
whether you bring them out with a plectrum
or pick out funeral songs in doleful complaints,
or popular rhythms. These Venus imbues
with a nectarous drink and threefold grace;
she is able to teach to drive out inertia;

1–2 Cf. *Theorica* 1.1.4a.6.
12 Cf. *Theorica* 1.1.78.
20–21 Cf. *Theorica* 1.1.73.
25 Cf. *Theorica* 1.1.248.
29 Cf. *Theorica* 1.1.129, 138
30–31 Cf. *Theorica* 1.1.9c.

whether the banquets or dances tempt you,
when the Virgin casts easy Ionic movements,
or when the Virgin, conscious of the strong passionate longings, half-hidden smiles
at a young man from the corner of her eye.
And he returns the words with silent grace.
When Cypris hurls the dart to his inmost heart,
then this false pertness, in which he is threatened
by the offended Virgin with her nails, is preserved in songs.
However music accords better with peace and is useful as a pastime,
for games. It supplies very dreadful war with an urging sound.
And music does not seize those standing firm
in the frightful critical moment. Such great rage is present
in the sounding trumpet; and there are those blowing on the clarion
who incite to battle and stimulate the minds to rush to arms and to the horses.
The trumpet gives new desires to souls; it revives; it drives away [the tiredness, enemies].
The wise city of Lacedaemon is aware of this,
although it rejected the many-stringed lyre.
Played in the middle of a banquet,
the trumpet induces the Macedonian man to leave the banquet, to take up arms.
But when the mode is changed to Phrygian, he soon returns to his couch.
Pythagoras, the priest of this music, has publicly declared
that it was possible to restrain drunken men
by blowing loudly the aulos in the Dorian mode [k6r].
You desire to know instruments which are diverse
and to know many by their sounds. Plato, Horace,
Vergil, the eloquent, Tullius [Cicero] as well as
the learned Fabius [Quintilian] and others asserted repeatedly
that the sacred origin of music stems from nature. The archaic

33 Cf. *Theorica* 1.1.101, 134–136 on the banquet; *Theorica* 1.1.212–217 on the dance.
41 Cf. *Theorica* 1.1.74, 101, 134, 137, 141, 213–214, 220–227; for a discussion of these passages, see Kreyszig, *Franchino Gaffurio's Theorica musice,* chapter 3, section 2.
42 Cf. *Theorica* 1.1.211a, 221 on games.
45–46 Cf. *Theorica* 1.1.108.
48 Cf. *Theorica* 1.1.106.
51–52 Cf. *Theorica* 1.1.80.
54 Cf. *Theorica* 1.1.125.
56–57 Cf. *Theorica* 1.1.2–4, 12–13, 15–20, 24–28, 30–34, 49–51, 228–2231; for a discussion of these passages, see Kreyszig, *Franchino Gaffurio's Theorica musice,* chapter 3, section 2.

art formerly maintained the laws of the Roman citizens.
If the art gave up its reserve slowly and was acknowledged seriously,
it did not make the learner ashamed to grow old in it.
And you, learned old man, have you adjusted
your unskilled fingers to try the sweet kithara?
Oh, he was not greatly ashamed to learn even as an old man,
because it was unseemly for a man not to know.
The powerful Miltiades studied it at night.
Let the Socratic work be like a mirror:
she [music, Terpsichore, etc.] unites the soft souls and moderates the old ones.
and has assisted the rulers with suitable care.
So, the old centaur Aeacides, living in a cave,
fixed his great hands to the kithara,
in which after the destruction of Troy
the death of Hector is made known to the Greeks with string music.
The furor sprung forth in the massacre by the melodies
just as the lyres sustain the one returning from the war.
After he had put on small wreaths of parsley,
the cruel emperor Nero sang to the lyre
during the contest of bloody killing.
The Babylonian king ordered the deaf image
to hear the hundredfold sound.
Moses, who has delivered the circumcised race from Pharaoh's yoke,
allowed music in the sanctuary.
In the holy domes, the organs alternate with voices.
The poor art is not fruitless labor;
the pleasant flock responds to that which the good Ethiopian
has recalled singing rhythms
favoring the inclinations of the ear,
since it is due to its powers that variety becomes unity.
The kings, the people, the little country, and the city
joyfully adore she who uplifts everything.
I do not know at all whether Sicilian Bronte and Sterope [k6v]
draw anything musical from the fiery anvil.

61 Cf. *Theorica* 1.1.233–234.
66–67 This is a reference to Socrates (cf. *Theorica* 1.1.259).
71 Cf. *Theorica* 1.1.95ff.
74 Cf. *Theorica* 1.1.9b.
79 Further on Nero's association with music, see. *Theorica* 1.1.97–98.
84 Cf. *Theorica* 1.1.236.
85 This statement is a reference to the widely known *alternatim* practice; see also *Theorica* 1.1.160.

The west winds blow and the globe is flaming,
but the leaping brook murmurs back delightfully.
She [i.e. music] solaces the raging sea which hungers
for the wreck of ships with its lashing waves,
and she comforts by melodies
and inveighs against the terrors of the hail
and raging south wind in their flow.
She restores whole strength to the army of farmers who cultivate the field
with a hooked ploughshare consisting of a light hoe or of a toothed hoe.
The pipe recommends itself to the flock and on the threshing-floor
it resounds more lowly to the voices of the reapers.
Because of the song the little girls spin thread
and do not feel the cold, the labor, and the weariness.
On a festive day, the singing shady elm
removes anxiety of work from the heart of the farmer.
Indeed, she is welcome to wild beasts;
here a lamb, a gora, and here a ram leaps.
And with trembling ears, the deer,
forgetting banquets, listens to the murmurs of the pipe,
while the hunter, lying in wait
on the tender green field of grass, calls out breaths with song.
The hare has trusted the trees of the wood
until he observed that they were moved by the Thracian strings.
And the singing bird who fills
the air with his songs, was astonished thus.
And it [i.e. the playing of the strings] soothens the meadows through her prophetical swans.
This is sweetly flowing milk through the gullet in nourishment.
And in the forests of the seabird, the male descendants of Cecrops sing,
and the parrot, which is rare in the rooms of the kings,
sings with loud voice in gilded cages.
Humming, also the bees return to their work

109 Cf. *Theorica* 1.1.148.
110 Cf. *Theorica* 1.1.63, 148, 152.
112–113 Cf. *Theorica* 1.1.68a.
118–119 Cf. *Theorica* 1.1.70–71.
120 Cf. *Theorica* 1.1.69.
125 Cf. *Theorica* 1.1.110–114.

because of the sweet honey, to the thymes on the impenetrable Hymettus.
And music is no annoyance to deaf fish.
Let witness the case of unfortunate Arion.
This music ravishes the fish, the rocks,
the smooth Tritonian sea and the friendly wet-haired Nereids.
Music can cause the always flowing rivers to stand still, [k7r]
and also it has pulled by melody the masses [k7r]
of immovable mountains down to moist valleys.
Nor did a stable river Dirce gape at the stones flying to Thebes,
when Amphion sang to the lyre.
Music even bears a pleasant work for the idle person.
And she, exhilarating the feeble, attracts leisure.
This brighter music, purer than ointments and thirsty amphoras,
provides the bridge for festive freedom.
This music drives away the burning anxiety.
And music lifts the troubled mind from fear.
Powerful is the assistance which music brings to the mind.
And music moves the dark cloud from the breast.
Because of music the region of the Galli is very famous,
known by the name as that the distinguished country of the Galli
is praised by the songs of the Galli, and mount Idea by tympani.
The band, tribe of Thessalians, like the Galli,
stamped their feet to the noise of the Berecyndians.
Music in the groves rouses up the goat-footed Satyr
and the half-god Faunus with his flock of Dryads.
And the horns of the swift Bacchus awaken the guard
who walks around on unsteady legs.
Is it astonishing if music soothes herds and the ridges of mountains,
and if it softens the lovely woods
with the rivers and seizes the sea and the wild beasts?

127 Cf. *Theorica* 1.1.61.
128 Cf. *Theorica* 1.1.65.
129 Cf. *Theorica* 1.1.57, 64.
132 Cf., for example, *Theorica* 1.1.84.
134–135 Cf. Hom. (*Od.* 11.2.260–265) and Hesiod (Frag. 96, ed. Evelyn-White). When Amphion and his twin brother Zetus built the walls of Thebes (cf. Hom. *Od.* 11.2.260–265), the stones set themselves in place through the power of the lyre (cf. Hesiod frag. 96); see Warren Anderson, "Amphion," *New Grove* 1:332; cf. *Theorica* 1.1.64.
140 Cf. *Theorica* 1.1.79, 224.
142 Cf. *Theorica* 1.1.75a.
153 Cf. *Theorica* 1.1.148.
155 Cf. *Theorica* 1.1.63.

Or is it surprising if music quiets the baby complaining at the breast,
or possesses the light minds of men?
Is it astonishing if Echo resounds in numerical proportions?
Or if the Siren holds hollow ships with her voice,
and the monstrous Cyclops solaces
the logs of his eagle eye on the unequal pipes,
and the Nicean stone has learned melodies?
Is it to be wondered that music hinders the violent rage of the demon?
Orcus that has no pity, listened to Orpheus
when, relying on his music, he pleaded for Eurydice.
After Orpheus had reconciled the alert Cerberus and the Styx,
the infernal boatman was amazed at them and at himself.
the angry Taenarceus, the father of the Eumenides,
sensed amid tears what piety meant [k7v].
Nothing elevates the souls more powerfully
or more strongly seizes on the ardent desire.
Music, more violent than the claws of the Harpies,
rejoices that she can achieve all things by songs.
I go forward to conquer the twisted Monycha,
never overcome by the notes of the Castilian Muses.
I think she will be moved, unless she should surpass the swollen ocean,
woods, rocks, beasts, the poison of Africa,
grim Persephone and gloomy Tartarus
in her enormous savages,
so that her unusual appearance rises above the dignity of Tyndaris.
Whatever the parent Nature covers with its starry sphere
and every man-made thing Music subdues.
Before music carried off a man from his father, who lay in want for him
This music makes Jupiter gentle, puts aside the Lemnian
thunderbolt, and yields to sung prayers.
The virgin, mother of the heavenly leader,
never touched with the light dew of desire, loved music,
meditating about both its origin and its pole.
And the poets moved twin-born Castor with musical
voices that he return the lights.
This music is the glory of Phoebus and in the choirs of the Pierian Muses.

156 Cf. *Theorica* 1.1.150, 199.
158 Cf. *Theorica* 2.2.26.
159 Further on the siren, see *Theorica* 1.1.162, 165.
163 Cf. *Theorica* 1.1.81.

This music is the glory of Phoebus, sweet in the banquets of the gods.
With the aid of this singing Diana halts in her course
Pan and Linus and swift Mercury;
wise Pallas, with her Aegis sought this music;
the Fates, the Clarian prophetic caves,
and the inspired voice of David sung of the true Jove
and the fates of nations with this sacred music.
Of all these fathers of music, Gaffurio shows himself a rival,
Franchino sweetly-speaking, with resounding melodies.
He who has made himself meritorious of this poem
is he who has restored the holy art of Calliope.
By his works music is revived.
May music flourish under his auspices.
May music come forth gleaming with his bronze.
He is the only one after the ancient writers who has written lengthy books
Singing he follows Boethius with his language [k8r].
In his life, his counsels, his art, his knowledge,
No one comes across as his equal in the trivium.
For this reason, beneficent with steady mind,
city, mistress of all affairs, city governing over all other cities
oh home of the gods, oh my celebrated native land,
oh fathers of the town, oh boys and you little girls!
Affirm that Franchino, with praises bestowed upon him,
is the grand culmination of the men belonging to Mercury!
Lift up Franchino with unrestrained honors
and gird Franchino with honorable garlands of leaves
and with myrtle or with the olive berry, or crown him
with twining ivy, or more so with the garlands of parsley,
or let Haemonian Daphne prepare a diadem
Daphne, who is properly the laurel of Apollo.
Let it bind the temples of the Thespician musician.
Praise be to God.

Printed in Milan by Master Filippo Mantegazzi, known as Castano, at the order and expense of Master Giovanni Petro de Lomazio in the year of our Savior 1492, on the fifteenth day of December.

194 On Linus, see *Theorica* 1.1.4–5; on Mercury, see *Theorica* 1.1.42.
197 Cf. *Theorica* 1.1.158.

EDITIONS AND TRANSLATIONS OF SOURCES CITED

Albertus Magnus: De quantitate

Borgnet, August, ed. *D. Alberti Magni: De quantitate* (*Tractatus III of Liber Primus: De praedicabilibus,* part of *Ordinis praedicatorum logica*). In August Borgnet, ed., *B. Alberti Magni Ratisbonensis Episcopi, Ordinis Praedicatorum, Opera Omnia,* pp. 194–221. Paris: Apud Ludovicum Vives, Bibliopolam Editorem, 1840–1899.

Alcaeus: Fragment 3

Bergk, Theodor, ed. *Poetas Melicos Continens.* Vol. 3 of *Poetae Lyrici Graeci.* Leipzig: B. G. Teubner, 1882, pp. 147–197.

Alcaeus: Fragment 307

Lobel, Edgar, and Enys Page, eds. *Poetarum Lesbiorum Fragmenta.* Oxford: Clarendon Press, 1955, pp. 112–259.

Alexander [Polyhistor]: Notices on Phrygia

Jacoby, Felix. *Geschichte von Staedten und Voelkern (Horographie und Ethnographie) A. Autoren ueber verschiedene Staedte,* vol. 3 of Felix Jacoby, *Die Fragmente der griechischen Historiker.* 1940; rpt. Leiden: E. Brill, 1954, no. 273 (pp. 96 ff).

Ambrose: Hexameron

Savage, John J., trans. *Saint Ambrose: Hexameron, Paradise, and Cain and Abel.* Vol. 42 of *The Fathers of the Church: A New Translation,* ed. Ludwig Schopp et al. New York: Fathers of the Church, 1961.

Anonymous: Musica Enchiriadis

Coussemaker, Edmond de, ed. *Hucbaldi Monachi Elnonensis quaedam e musica enchiriade inedita,* pp. 74–77 of vol. 2 of *Scriptorum de musica medii aevii novam seriem a Gerbertiana alteram.* 1867; rpt. Hildesheim: Georg Olms, 1963.

Gerbert, Martin, ed. *Hucbaldi Monachi Elonensis: Musica enchiriadis,* pp. 152–212 of vol. 1 of *Scriptores ecclesiastici de musica sacra potissimum ex variis Italiae, Galliae et Germaniae codicibus manuscriptis collecti et nunc primum publica luce donati.* 1784; rpt. Hildesheim: Georg Olms, 1963.

Schmid, Hans, ed. *Musica et Scolia enchiriadis: una cum aliquibus tractatulis adiunctis: Recensio nova post Gerbertinam altera ad fidem omnium codicum manuscriptorum,* vol. 3 of *Veröffentlichungen der Musikhistorischen Kommission.* Munich: Verlag der Bayerischen Akademie der Wissenschaften: (In Kommission bei der C. H. Beck'schen Verlagsbuchhandlung), 1981.

Anonymous XI: Tractatus de musica plana et mensurabili

Coussemaker, Edmond de., ed. *Anonymi XI: Tractatus de musica plana et mensurabili,* pp. 416–475 of vol. 3 of *Scriptorum de musica medii aevii novam seriem a Gerbertiana alteram.* 1867; rpt. Hildesheim: Georg Olms, 1963.

Wingell, Richard Joseph, ed. and trans. "Anonymous XI, (CS III): An Edition, Translation, and Commentary." 3 vols. Ph.D. diss., University of Southern California, 1973.

Anselmi, Giorgio: De musica

Massera, Giuseppe, ed. *Giorgii Anselmi Parmensis: De musica dieta prima de celesti harmonia, dieta secunda de instrumentali harmonia, dieta tertia de cantabili harmonia,* vol. 14 of *Historiae Musicae Cultores Biblioteca.* Florence: Leo S. Olschki, 1961.

Apuleius: Florida

Vallette, Paul, ed. and trans. *Apuleius Madaurensis: Apologie, Florides,* part of *Collection des Universités de France.* 1924; rpt. Paris: Les Belles Lettres, 1971.

Aristides Quintilianus: De musica

Burana, Giovanni Francesco, trans. *Musica e graeco in latinum conversa.* Verona, Biblioteca capitolare, MS CCXL (201), fols. 1 recto–25 verso, dated 15 April 1494.

Winnington-Ingram, R. P., ed. *Aristides Quintilianus: De musica libri tres,* part of *Bibliotheca scriptorum Graecorum et Romanorum Teubneriana.* Leipzig: B. G. Teubner, 1963.

Barker, Andrew, trans. "Aristides Quintilianus." In *Harmonic and Acoustic Theory,* pp. 392–535. Vol. 2 of *Greek Musical Writings,* ed. Andrew Barker. Cambridge Readings in the Literature of Music. Cambridge: Cambridge University Press, 1989.

Mathiesen, Thomas J., trans. *Aristides Quintilianus: On Music In Three Books.* Music Theory Translation Series,. New Haven: Yale University Press, 1983.

Aristophanes: Nubes (Clouds)

Rogers, Benjamin Bickley, ed. and trans. *Aristophanes. Loeb Classical Library.* 1967. 1:262–401.

Aristophanes: Rana (Frogs)

Rogers, Benjamin Bickley, ed. and trans. *Aristophanes. Loeb Classical Library.* 1967. 2:296–437.

Aristotle: Categoria

Boethius, trans. *Aristoteles Latinus: Categoriae vel Praedicamenta,* ed. Lorenzo Minio-Paluello, as part of *Corpus Philosophorum Medii Aevi Academiarum Consociatarum.* Bruges-Paris: Desclée de Brouwer, 1961.

Cooke, Harold P. ed. and trans. *Aristotle: The Organon-the Categories, On Interpretaton. Loeb Classical Library.* 1955.

Aristotle: De anima. See under: Themistius.

Aristotle: De caelo

Allan, Donald James, ed. *Aristotelis De caelo libri quattuor,* part of *Scriptorum classicorum bibliotheca oxoniensis.* 1936; rpt. Oxford: Clarendon, 1955.

Guthrie, William Keith Chambers, ed. and trans. *Aristotle: On the Heavens. Loeb Classical Library.* 1939.

Aristotle: Ethica Nicomachea

Rackham, H., ed. and trans. *Aristotle: The Nicomachean Ethics. Loeb Classical Library.* 1926.

Aristotle: Historia Animalium

Peck, A. L., ed. and trans. *Aristotle: Historia Animalia. Loeb Classical Library.* 1965–1970.

Aristotle: Metaphysica

Tredennick, Hugh, ed. and trans. *Aristotle: The Metaphysics,* 2 vols. *Loeb Classical Library.* 1956–1962.

Aristotle: Physica

Wicksteed, Philip and Francis M. Cornford, ed. and trans. *Aristotle: The Physics.* 2 vols. *Loeb Classical Library.* 1935–1957.

Aristotle: Poetica

Fyfe, W. Hamilton, ed. and trans. *Aristotle: The Poetics. Loeb Classical Library.* 1932, pp. 1–118.

Aristotle: Politica

Rackham, H., ed. and trans. *Aristotle: Politics. Loeb Classical Library.* 1959.

Aristotle: Problemata

d'Abano, Pietro. *Expositio problematum Aristotelis (cum textu latine).* Mantua: Paulus Johannis de Puzpach, 1475.

Foster, E. S., trans. *Aristotle: Problems.* Vol. 7 of *The Works of Aristotle,* ed. W. D. Ross. Oxford: Clarendon Press, 1927.

Barker, Andrew, trans. "The Aristotelian Problems."In *The Musician and His Art,* pp. 190–204. Vol. 1 of *Greek Musical Writings,* ed. Andrew Barker. *Cambridge Readings in the Literature of Music.* Cambridge: Cambridge University Press, 1984.

Barker, Andrew, trans. "The Aristotelian *Problemata.*" In *Harmonic and Acoustic Theory,* pp. 85–97. Vol. 2 of *Greek Musical Writings,* ed. Andrew Barker. *Cambridge Readings in the Literature of Music.* Cambridge: Cambridge University Press, 1989.

Aristotle: Topica

Forster, E. S., ed. and trans. *Aristotle: Topics. Loeb Classical Library.* 1960, pp. 265–739.

Stump, Eleanore, trans. *Boethius's De topicis differentiis.* Ithaca: Cornell University Press, 1978.

Aristoxenus: Harmonica

Macran, Henry S., ed. and trans. *The Harmonics of Aristoxenus.* 1902; rpt. Hildesheim: Georg Olms, 1974.

Barker, Andrew, trans. "Aristoxenus." In *Harmonic and Acoustic Theory,* pp. 119–189. Vol. 2 of *Greek Musical Writings,* ed. Andrew Barker. *Cambridge Readings in the Literature of Music.* Cambridge: Cambridge University Press, 1989.

Athenaeus: Deipnosophistai

Gulick, Charles Burton, ed. and trans. *Athenaeus: The Deipnosophists.* 6 vols. *Loeb Classical Library.* 1927–1941.

Augustine of Hippo, Aurelius: De civitate Dei

Augustine: De civitate Dei. Venice: Johann and Wendelin of Speyer, 1470; Venice: Gabriele die Pietro, 1475; Venice: Nicolaus Jensen, 2 October 1475; Basel: Thomas Valois and Nicolai Triveth, 1489; Venice: Boretus Locatellus for Octavianus Scotus, 18 February 1489/90; Basel: Johann Amerbach, 1490.

McCracken, G. E., et al., trans. and ed. *Augustine: The City of God Against the Pagans. Loeb Classical Library.* 1957–72.

Augustine: De musica

Marzi, Giovanni, trans. and ed. *Aurelii Augustini: De musica.* Vol. 1 of *Collana di classici della filosofia cristiana,* ed. by Carlo Giacon. Florence: Sansoni, 1969.

Taliaferro, Robert Catesby, trans. *Saint Augustine: On Music,* pp. 153–379. Vol. 2 of *Writings of Saint Augustine.* Vol. 5 of *The Fathers of the Church: A New Translation.* New York: Cima Publishing, 1947.

Bacchius of Geron: Introductio artis musicae

Bellermann, Friedrich, ed. *Anonymi scriptio de musica Bachii senioris introductio artis musicae.* Berlin, 1841.

Steinmayer, Otto, trans. "Bacchius Geron's *Introduction to the Art of Music,*" *Journal of Music Theory* 29, no. 2 (1985), pp. 271–298.

Barbaro, Ermolao. See under: Themistius.

Boethius, Anicius Manlius Severinus: De institutione arithmetica and De institutione musica

Boethius: Opera. Venice: Giovanni e Gregorio de Gregori, 18 August 1492.

Friedlein, Gottfried, ed. *Anicii Manlii Torquati Severini Boetii: Institutione arithmetica libri duo e Institutione musica libri quinque accedit Geometria quae fertur Boetii.* 1867; rpt. Frankfurt am Main: Minerva, 1966.

Masi, Michael. *Boethian Number Theory: A Translation of the "De institutione arithmetica",* vol. 6 of *Studies in Classical Antiquity.* Amsterdam: Rodopi, 1983.

Bower, Calvin, trans. Anicius Manlius Severinus Boethius, *Fundamentals of Music.* Music Theory Translation Series. New Haven: Yale University Press, 1989.

Bryennius, Manuel: Harmonica

Jonker, G. H., ed. and trans. *The Harmonics of Manuel Bryennius.* Groningen: Walters-Noordhoff Publishing, 1970.

Cassiodorus Senator, Flavius Magnus Aurelius: Variae

Fridh, A. J., ed. *Cassiodorus: Variarum libri XII.* Turnholt: Typographic Brepols Editores Pontificii, 1973.

Cassiodorus: Institutiones humanarum litterarum

Goode, Helen Dill, and Gertrude C. Drake, trans. *Institutiones humanarum litterarum.* Vol. 12 of *Colorado College Music Press.* Colorado Springs: Colorado College Music Press, 1980.

Cato, Marcus Porcius: De agricultura

Hooper, William Davis, ed. and trans. *Marcus Porcius Cato: On Agriculture.* rev. ed. *Loeb Classical Library.* 1967, pp. 1–157.

Censorinus: De die natali

Hultsch, Friedrich, ed. *Censorini: De die natali liber.*Leipzig: B. G. Teubner, 1867.

Maude, William, trans. *De die natali ("The natal day") by Censorinus (A.D. 238) Life of the Emperor Hadrian by Helianus Spartianus (Circ. A.D. 300).* New York: Cambridge Encyclopedia, 1900.

Chalcidius: In Platonis Timaeum

Waszink, J. H., ed. *Timaeus a Calcidio translatus commentarioque instructus.* Vol. 4 of *Plato Latinus,* ed. Raymund Klibansky, as part of *Corpus Platonicum Medii Aevi.* London: Warburg Institute and Leiden: E. J. Brill, 1962.

Cicero, Marcus Tullius: Ad de ratione dicendi C. Herennium

Caplan, Harry, ed. and trans. *Ad C. Herennium--De ratione dicendi.* Vol. 1 of *Cicero,* 28 vols. *Loeb Classical Library.* 1954.

Cicero: De legibus

Keyes, Clinton Walker. *Cicero: De legibus. Loeb Classical Library.* 1977, pp. 287–519.

Cicero: De oratore

Sutton, E. W., ed. and trans. *Cicero: De oratore*. 2 vols. *Loeb Classical Library*. 1942.

Cicero: De re publica

Keyes, Clinton Walker, ed. and trans. *Cicero: De Re Publica*. *Loeb Classical Library*. 1977, pp. 1–285.

Cicero: Tusculanae disputationes

Pohlenz, M., ed. *Tulli Ciceronis Tusculanae disputationes*. Vol. 44 of *Bibliotheca scriptorum graecorum et romanorum Teubneriana*. Stuttgart: B. G. Teubner, 1967.

Cleonides: Introduction to Harmonics

Valla, Giorgio, trans. *Cleonides: Harmonicum introductorium*. Venice: Simon Papiens dictus Bevilaqua, 1497.

Ruelle, Charles Emile, trans. *L'introduction harmonique de Cléonide, la division du canon d'Euclide le géometre, Canons harmoniques de Florence*. Vol. 3 of *Collection des auteurs grecs relatifs à la musique*. Paris, 1884.

Conrad von Zabern: Novellus musicae artis tractatus

Gümpel, Karl-Werner. *Die Musiktraktate Conrads von Zabern*. Vol. 4 of *Akademie der Wissenschaften und der Literatur: Abhandlungen der geistes- und sozialwissenschaftlichen Klasse 1956*. Mainz: Verlag der Akademie der Wissenschaften und der Literatur and Wiesbaden: Franz Steiner Verlag, 1956, pp. 40–100.

Democritus of Abdera: On Rhythms and Harmony (fragm.)

Diels, Hermann. *Die Fragmente der Vorsokratiker*. 3 vols., ed. W. Kranz. Berlin-Grunewald: Weidmannsche Verlagsbuchhandlung, 1951–1952, 2:130–207.

Diogenes Laertius: Vitae Philosophorum

Hicks, R. D., ed. and trans. *Diogenes Laertius: Lives of Eminent Philosophers*. 2 vols. *Loeb Classical Library*. 1942–1950.

Dionysius of Halicarnassus: Antiquitates Romanae

Cary, Earnest, ed. and trans. *The Roman Antiquities of Dionysius of Halicarnassus*. 7 vols. *Loeb Classical Library*. 1937–1939.

Euripides: Orestes

Biehl, Werner, ed. *Euripides: Orestes. Bibliotheca Scriptorum graecorum et romanorum Teubneriana,* ed. Akademie der Wissenschaften der DDR: Zentralinstitut für Geschichte und Archäologie. Leipzig: B. G. Teubner, 1975.

Willink, C. W., ed. *Euripides: Orestes.* Oxford: Clarendon Press, 1986.

Fabius Quintilian: Istitutio oratoria

Butler, H. E., ed. and trans. *The Institutio Oratoria of Quintilian.* 2 vols. *Loeb Classical Library.* 1921.

Ficino, Marsilio. See under: Plato.

Finck, Hermann: Practica musica

Matzdorf, Paul. *Die Practica musica Hermann Fincks.* Frankfurt am Main, 1957.

Gaffurio, Franchino: De harmonia musicorum instrumentorum opus

Gaffurio, Franchino. *De harmonia musicorum instrumentorum opus.* Milan: Gotardus Pontanus Calographus, 27 November 1518.

Vecchi, Giuseppe. Facsimile Edition of *Franchino Gaffurio: De harmonia musicorum instrumentorum opus.* Vol. 2/7 of *Bibliotheca musica Bononiensis.* Bologna, 1972.

Miller, Clement A., trans. *Franchinus Gaffurius: De harmonia musicorum instrumentorum opus.* Vol. 33 of *Musicological Studies and Documents.* American Institute of Musicology, 1977.

Gaffurio: Practica musicae

Franchino Gaffurio: Practica musice. Milan: Guilermus Signerre, 1496. Republished as facsimile edition. Farnborough: Gregg Press, 1967.

Miller, Clement A., trans. *Franchinus Gaffurius: Practica musicae.* Vol. 20 of *Musicological Studies and Documents.* American Institute of Musicology, 1968.

Vecchi, Guiseppe. Facsimile Edition of *Franchino Gaffurio: Practica musicae.* Vol. 2/7 of *Bibliotheca musica Bononiensis.* Bologna, 1972.

Young, Irwin, trans. *The Practica musicae of Franchinus Gaffurius.* Madison: University of Wisconsin Press, 1969.

Gaffurio: Theorica musice

Cesari, Gaetano. Facsimile Edition of *Franchino Gaffurio: Theorica musice.* Rome, 1934.

Franchino Gaffurio: Theorica musice (A Facsimile of the 1492 Milan Edition). Vol. 21 of *Monuments of Music and Music in Literature in Facsimile—Second Series: Music Literature*. New York: Broude Brothers, 1967.

Vecchi, Guiseppe. Facsimile Edition of *Franchino Gaffurio: Theorica musice*. Vol. 2/5 of *Bibliotheca musica Bononiensis*. Bologna, 1969.

Gallicus, Johannes: Libelli musicales de ritu canendi vetustissimo et novo

Coussemaker, Edmond de, ed. *Joannis Gallici dicti Carthusiensis seu de Mantua: Ritus canendi vetustissimus et novus*, pp. 298–421. Vol. 4 of *Scriptorum de musica medii aevii novam seriem a Gerbertiana alteram*. 1876; rpt. Hildesheim: Georg Olms, 1963.

Albert Seay, ed. *Johannes Gallicus Legrense: Ritus canendi vetustissimus et novus*, part of *Colorado College Music Press Series*. Colorado Springs: Colorado College Music Press, 1981.

Gellius, Aulus: Noctes Atticae

Rolfe, John Carew, ed. and trans. *The Attic Nights of Aulus Gellius*. 3 vols. *Loeb Classical Library*. 1969–71.

Guido of Arezzo: Epistola de ignotu cantu

Gerbert, Martin, ed. *Epistola Guidonis Michaeli Monacho: De ignoto cantu*, pp. 43–50. Vol. 2 of *Scriptores ecclesiastici de musica sacra potissimum ex variis Italiae, Galliae et Germaniae codicibus manuscriptis collecti et nunc primum publica luce donati*. 1784; rpt. Milan: Bollettino Bibliografico Musicale, 1931.

Smits van Waesberghe, Joseph, ed. *Guidonis Aretini tres tractatuli editi cum apparatu critico: Epistola ad Michaelem*. Vol. 5 of *Divitiae Musicae Artis (Series A)*. Buren: Frits Knuf, 1976.

Guido of Arezzo: Epistola ad Michaelem

Gerbert, Martin, ed. *Epistola Guidonis Michaeli Monacho: De ignoto cantu*, pp. 43–50. Vol. 2 of *Scriptores ecclesiastici de musica sacra potissimum ex variis Italiae, Galliae et Germaniae codicibus manuscriptis collecti et nunc primum publica luce donati*. 1784; rpt. Milan: Bollettino Bibliografico Musicale, 1931.

Guido of Arezzo: Micrologus

Gerbert, Martin, ed. *Micrologus Guidonis: De disciplina artis musicae*, pp. 2–24. Vol. 2 of *Scriptores ecclesiastici de musica sacra potissimum ex variis Italiae, Galliae et Germaniae codicibus manuscriptis collecti et nunc primum publica luce donati*. 1784; rpt. Milan: Bollettino Bibliografico Musicale, 1931.

Smits van Waesberghe, Joseph, ed. *Guidonis Aretini Micrologus*. Vol. 4 of *Corpus Scriptorum de Musica,* ed. Armen Carapetyan. American Institute of Musicology, 1955.

Babb, Warren, and Claude V. Palisca, ed.. *Hucbald, Guido, and John on Music: Three Medieval Treatises*. Vol. 3 of Music Theory Translation Series. New Haven: Yale University Press, 1978.

Guido of Arezzo: Regulae rhythmicae

Gerbert, Martin, ed. *Musicae Guidonis regulae rhythmicae,* pp. 25–34. Vol. 2 of *Scriptores ecclesiastici de musica sacra potissimum ex variis Italiae, Galliae et Germaniae codicibus manuscriptis collecti et nunc primum publica luce donati.* 1784; rpt. Milan: Bollettino Bibliografico Musicale, 1931.

Smits van Waesberghe, Joseph, ed. *Guidonis Aretini tres tractatuli editi cum apparatu critico: Regulae dictae Rhythmicae,* vol. 4 of *Divitiae Musicae Artis (Series A).* Buren: Frits Knuf, 1985.

Herodotus: Natural History

Goodley, A. D., ed. and trans. *Herodotus*. 4 vols. *Loeb Classical Library. 1920–1949.*

Hesiod

Evelyn-White, Hugh G., ed. and trans. *Hesiod, the Homeric Hymns, and Homerica. Loeb Classical Library*. 1920.

Hieronymous (Jerome): Dialogus contra Pelagianos

Hritzu, John N., trans. *Saint Jerome: Dogmatic and Polemic Works*. Vol. 53 of *The Fathers of the Church: A New Translation*. Washington, D.C.: Catholic University of America Press, 1965, pp. 223–378.

Homer: The Iliad

Murray, A. T., ed. and trans. *Homer: The Iliad*. 2 vols. *Loeb Classical Library*. 1960.

Homer: The Odyssey

Murray, A. T., ed. and trans. *Homer: The Odyssey*. 2 vols. *Loeb Classical Library*. 1969.

Horace, Flaccus Quintus: Ars poetica

Dunsany, Lord, and Michael Oakley, trans. *The Collected Works of Horace*. Everyman's Library, 1967.

Hucbald: De musica

Gerbert, Martin, ed. *Ubaldi seu Hucbaldi Monachi Elnonensis opuscula De musica,* pp. 103–152. Vol. 1 of *Scriptores ecclesiastici de musica sacra potissimum ex variis Italiae, Galliae et Germaniae codicibus manuscriptis collecti et nunc primum publica luce donati*. 1784; rpt. Hildesheim: Georg Olms, 1963.

Babb, Warren, trans. *Hucbald: Melodic Instruction (De harmonica institutione)*. In Warren Babb, trans. *Hucbald, Guido, and John On Music: Three Medieval Treatises,* pp. 3–46. Vol. 3 of Music Theory Translation Series. New Haven: Yale University Press, 1978.

Isidore of Seville: Etymologiae sive origines

Lindsay, Wallace Martin, ed. *Isidori Hispalensis episcopi Etymologiarum sive originum libri XX,* part of *Scriptorum classicorum bibliotheca Oxoniensis*. 1911; rpt. Oxford: Clarendon Press, 1962.

Jacques de Liège: Speculum musicae

Bragard, Roger, ed. *Jacobi Leodiensis: Speculum Musicae*. Vol. 3 of *Corpus Scriptorum de Musica,* ed. Armen Carapetyan. 7 vols. American Institute of Musicology, 1955–1973.

Josephus, Flavius: Antiquitates Judaicae

Thackeray, H. St. J., ed. and trans. *Josephus: Jewish Antiquities*. 8 vols. *Loeb Classical Library*. 1930–1963.

Legrense, Johannes. See under: Gallicus.

Livy Andronicus, Lucius: Opera omnia

Foster, Benjamin Oliver, ed. and trans. *Livy*. 14 vols. *Loeb Classical Library*. 1922–1959.

Lucian of Samosata: Dialogi Deorum

MacLeod, M. D. et al., ed. and trans. *Lucian*. 8 vols. *Loeb Classical Library*. 1921–1967, 7(1961):239–353.

Lucretius, Titus Carus: De rerum naturae

Bailey, Cyril, trans. and ed. *Titi Lucreti Cari: De rerum natura libri sex*. 2 vols. Oxford: Clarendon Press, 1947.

Macrobius, Ambrosius Aurelus Theodosius: Commentarius in Ciceronis Somnium Scipionis

Willis, Jacob, ed. *Saturnalia, apparatu critico instruxit, in Somnium Scipionis commentarios selecta varietate lectionis*. 2nd. ed., part of *Biblioteca scriptorum graecorum et romanorum Teubneriana*. Leipzig: B. G. Teubner, 1970.

Marchetto of Padua: Lucidarium

Gerbert, Martin, ed. *Marcheti de Padua Musica seu Lucidarium in arte musicae planae ex Msc. Biblioth. Ambros. Mediolanensis,* pp. 64–121. Vol. 3 of *Scriptores ecclesiastici de musica sacra*. 1784; rpt. Hildesheim: Georg Olms, 1963.

Herlinger, Jan W. *The Lucidarium of Marchetto of Padua* (A Critical Edition, Translation, and Commentary). Chicago: University of Chicago Press, 1985.

Martial, Marcus Valerius: Epigrammes

Ker, W. C. A., ed. and trans. *Martial: Epigrams*. 2 vols. *Loeb Classical Library*. 1925.

Martianus Capella: De nuptiis Philologiae et Mercurii

Dick, Adolf, ed. *Martianus Capella,* part of *Bibliotheca scriptorum graecorum et romanorum Teubneriana*. 1925; rpt. Stuttgart: B. G. Teubner, 1969.

Stahl, W. H., and R. Johnson with E. L. Burge, trans. *Martianus Capella and the Seven Liberal Arts*. 2 vols. New York: Columbia University Press, 1977.

Nicetus: De laude et utilitate

Burn, Andrew Eubank. *Niceta of Remesiana: His Life and Works*. Cambridge: Cambridge University Press, 1905. (Includes edition of treatise).

Nicomachus of Gerasa: Introduction to Arithmetic

D'Ooge, Martin Luther, trans. *Introduction to Arithmetic, with Studies in Greek Arithmetic* by Frank Egelston Robbins and Louis Charles Karpinski. Vol. 16 of *University of Michigan Humanistic Series*. 1926; rpt. New York: Johnson Reprint, 1972.

Johnson, George. *The Arithmetic Philosophy of Nicomachus of Gerasa*. Lancaster, Pa.: Press of the New Era Printing Company, 1916. (Includes partial translation).

Nicomachus of Gerasa: Manual of Harmonics

Levin, Flora Rose. "Nicomachus of Gerasa: Manual of Harmonics (Translation and Commentary)." Ph.D. diss., Columbia University, 1967.

Barker, Andrew, trans. "Nicomachus." In *Harmonic and Acoustic Theory*, pp. 245–269. Vol. 2 of *Greek Musical Writings*, ed. Andrew Barker. *Cambridge Readings in the Literature of Music*. Cambridge: Cambridge University Press, 1989.

Nonnus of Panopolis: Dionysiaca

Rouse, W. H. D., ed. and trans. *Nonnus Panopolitanus: Dionysiaca. Loeb Classical Library*. 1940–42.

Ovid, Publius Naso: Fasti

Marsi, Pauli, ed. *Ovidius Naso Publius: Fasti*. Venice: Baptista de Tartis, 24 December 1482. Milan: Antonius Zarotus for Johannes de Legnano, 5 June 1483.

Frazer, Sir James George, ed. and trans. *Ovid's Fasti. Loeb Classical Library*. 1951.

Ovid: Metamorphoses

Miller, Frank Justus, ed. and trans. *Ovid: Metamorphoses*. 2 vols. *Loeb Classical Library*. 1916–1946.

Pausanias of Lydia: Descriptis Graeciae

Jones, W. H. S., ed. and trans. *Pausanias: Description of Greece*. 2 vols. *Loeb Classical Library*. 1931–1935.

Plato: Epinomis

Ficino, Marsilio, trans. *Opera*. Basel: Officina of Heinrich Petri, 1576.

Harward, J., trans. *The Epinomis of Plato*. Oxford: Clarendon Press, 1928.

Plato: Leges

Ficino, Marsilio, trans. *Opera.* Basel: Officina of Heinrich Petri, 1576.
Bury, Robert Gregg, ed. and trans. *Plato: Laws. Loeb Classical Library.* 1926.

Plato: Respublica

Ficino, Marsilio, trans. *Opera.* Basel: Officina of Heinrich Petri, 1576.
Shorey, Paul, ed. and trans. *Plato: The Republic,* 2 vols. *Loeb Classical Library.* 1930; rpt. : 1969.

Plato: Phaedrus

Rowe, C. J., ed. and trans. *Plato: Phaedrus.* 1986; 2nd. ed., Warminster, Wilts., England: Aris and Phillips, 1988.

Plato: Politica

Skemp, J. B. *Plato's Statesman* (A Translation of the *Politicus* of Plato with Introductory Essays and Footnotes). 1952; 2nd ed. Bristol: Bristol Classical Press, 1987.

Plato: Sophistes

Fowler, Harold North, ed. and trans. *Plato: Sophist. Loeb Classical Library.* 1961, pp. 259–459.

Plato: Timaeus

Ficino, Marsilio, trans. *Opera.* Basel: Officina of Heinrich Petri, 1576.
Lee, Desmond, trans. *Plato: Timaeus and Critias.* Baltimore: Penguin Books, 1971.
Kristeller, Paul, ed. *Marsilio Ficino: Opera Omnia.* Series I. Vols. 7–8 of *Monumenta Politica et Philosophica Rariora,* ed. by Luigi Firpo. Turin: Bottega d'Erasmo, 1962.

Pliny Senior, Gaius Secundus: Historia Naturalis

Jones, W. H. S., ed. and trans. *Pliny: Natural History (Libri XXVIII–XXXII). Loeb Classical Library.* 1975.
Rackham, H., ed. and trans. *Pliny: Natural History (Libri VIII–XI). Loeb Classical Library.* 1940.

Pliny Junior, Gaius Caecilius Secundus: Epistulae

Radice, Betty, ed. and trans. *Pliny: Letters and Panegyricus.* 2 vols. *Loeb Classical Library.* 1949.

Plutarch, Mestrius: De animae procreatione in Timaeo (On the Generation of the Soul in the Timaeus)

Cherniss, Harold, ed. and trans. *Plutarch's Moralia. Loeb Classical Library.* 1927–1976, 13(1976):131–345.

Plutarch: De musica

Einardson, Benedict, and Phillip H. DeLacy, ed. and trans. *Plutarch's Moralia. Loeb Classical Library.* 1927–1967, 14(1967):352–455.

Barker, Andrew, trans. "The Plutarchian treatise *On Music.*" In *The Musician and his Art,* pp. 205–248. Vol. 1 of *Greek Musical Writings,* ed. Andrew Barker. *Cambridge Readings in the Literature of Music.* Cambridge: Cambridge University Press, 1984.

Plutarch: De vita Alcibiadis

Perrin, Bernadotte, ed. and trans. *Alcibiades and Coriolanus; Lysander and Sulla.* Vol. 4 of *Plutarch's Lives. Loeb Classical Library.* 1916; rpt.: 1959, pp. 2-115.

Porphyry: Harmonica

Düring, Ingemar. *Ptolemaios und Porphyrios Über die Musik.* Vol. 40, No. 1 of *Göteborgs Högskolas Ārsskrift.* Göteborg: Elanders Boktryckeri Aktiebolag, 1934.

Proclus: Chrestomathia

Sveryns, Albert. *Recherches sur la Chrestomathie de Proclos,* 4 vols. Liège: Faculté de Philosophie et Lettres, 1938–1963. (Includes edition and French translation).

Prosdocimo de' Beldomandi: Parvus tractatulus de modo monacordum dividendi

Herlinger, Jan. *Prosdocimo de' Beldomandi: Brevis Summula Proportionum Quantum ad Musicam Pertinet and Parvus Tractatulus de Modo Monacordum Dividendi (A Short Summary of Ratios Insofar as They Pertain to Music and Little Treatise on the Method of Dividing the Monochord)* part of *Greek and Latin Music Theory.* Lincoln: University of Nebraska Press, 1987, pp. 64–118.

Ptolemy, Claudius: Harmonica

Claudius Ptolemaeus: Harmonicorum Libri Tres (A Facsimile of the Oxford 1682 Edition). Vol. 60 of *Monuments of Music and Music Literature in Facsimile: Second Series–Music Literature*. New York: Broude Brothers, 1977.

Barker, Andrew, trans. "Ptolemy." In *Harmonic and Acoustic Theory*, pp. 270–391. Vol. 2 of *Greek Musical Writings*, ed. Andrew Barker. Cambridge Readings in the Literature of Music. Cambridge: Cambridge University Press, 1989.

Düring, Ingemar. *Die Harmonielehre des Klaudios Ptolemaios*. Vol. 38, No. 2 of *Göteborgs Högskolas Årsskrift*. Göteborg: Elanders Boktryckeri Aktiebolag, 1930.

Strabo: Geographica

Jones, Horace Leonard, ed. and trans. *The Geography of Strabo*. 8 vols. *Loeb Classical Library*. 1917–1932.

Suetonius Tranquillus, Gaius: Nero

Warmington, B. H., ed. *Suetonicus: Nero*. Bristol: Bristol Classical Press, 1977.

Themistius: Paraphrases on the De anima of Aristotle

Moerbeke, William of, trans. *Themistius: Paraphrases on the De anima of Aristotle*, edited by G. Verbeke, as *Themistius: Commentaire sur le Traité de l'Ame d'Aristote*. Vol. 1 of *Centre de Wulf-Mansion Corpus Latinum Commentariorum in Aristotelem Graecorum*. Paris: Editions Beatrice-Nauwelaerts and Louvain: Publications Universitaires de Louvain, 1957.

Heinze, Richard, ed. *Paraphraseos de anima libri tres*. Trans. Ermolao Barbaro. Berlin, 1899.

Barbaro, Ermolao, trans. *Paraphrasis in Aristotelem*. Treviso: Bartolomeo Confalomeri and Morello Gerardino, 15 February 1481.

Tinctoris, Johannes: De inventione et usu musicae

Weinmann, Karl. *Johannes Tinctoris (1445–1511) und sein unbekannter Traktat 'De inventione et usu musicae': Historisch-kritische Untersuchung*, ed. Wilhelm Fischer. Tutzing: Hans Schneider, 1961.

Woodley, Ronald. "The Printing and Scope of Tinctoris's Fragmentary Treatise 'De inventione et usu musice.'" In *Studies in Medieval and Early Modern Music*, ed. Iain Fenlon, pp. 239–268. Vol. 5 of Early Music History. Cambridge: Cambridge University Press, 1985. (Includes partial edition).

Tinctoris: Expositio manus

Coussemaker, Edmond de. *Expositio manus secundum Magistrum Johannem Tinctoris in legibus licentiatum ac regis Sicille Capellanum,* pp. 1–200. Vol. 4 of *Scriptorum de musica medii aevi novam seriem a Gerbertiana alteram.* 1867; rpt. Hildesheim: Georg Olms, 1963

Seay, Albert. "The Expositio manus of Johannes Tinctoris." *Journal of Music Theory* 9 (1965), pp. 194–232.

Tinctoris: Proportionale Musices

Coussemaker, Edmond de., ed. *Proportionale musices editum a Magistro Joanne Tinctoris in legibus licenciatio serendissimique principis Ferdinandei regis Siciliae, Jherusalem et Ungariae cappellani feliciter,* pp. 153–177. Vol. 4 of *Scriptorum de musica medii aevi novam seriem a Gerbertiana alteram.* 1867; rpt. Hildesheim: Georg Olms, 1963.

Seay, Albert. "The *Proportionale Musices* of Johannes Tinctoris." *Journal of Music Theory* 1 (1957), pp. 22–75.

Seay, Albert, trans. *Johannes Tinctoris: Proportions in Music (Proportionale Musices).* Vol. 10 of *Colorado College Music Press Translations.* Colorado Springs: Colorado College Music Press, 1979.

Ugolino of Orvieto: Declaratio musicae disciplinae

Seay, Albert, ed. *Ugolino Urbevetanis: Declaratio musicae disciplinae* (3 parts). Vol. 7 of *Corpus Scriptorum de Musica,* ed. Armen Carapetyan. American Institute of Musicology, 1956.

Valerius Maximus

Kempf, Carl, ed. *Valerii Maximi Factorum et dictorum memorabilium libri novem cum Iulii Paridis et Ian varii Neporiani epitomis iterum.* Stuttgart, 1966.

Varro, Marcus Terentius: De rerum rustica

Hooper, William Davis, ed. and trans. *Marcus Terentius Varro: On Agriculture.*Rev. ed. *Loeb Classical Library.* 1967, pp. 159–529.

Virgil, Publius Maro: Aeneid

Page, T. E., ed. *The Aeneid of Virgil (Books 1–12).* 2 vols, part of Classical Series. London: Macmillan, 1964.

Knight, William Francis Jackson, trans. *The Aeneid by Virgil.* Vol. 51 of Penguin Classics. Harmondsworth: Penguin Books, 1960.

Virgil: Eclogues

Williams, R. D., ed. *The Eclogues and Georgics by Virgil,* part of *Classical Series.* New York: St. Martin's Press, 1979.

Trevelyan, R. C., trans. *Virgil: The Eclogues and Georgics.* Cambridge: Cambridge University Press, 1944. (English verse translation).

Virgil: Georgics. See under: Virgil: Eclogues.

Vitruvius Pollio, Marius: De architectura

Granger, Frank, ed. and trans. *Vitruvius, On architecture.* (Harleian Manuscript 27.67). *Loeb Classical Library,* 1931–1934.

INDEX

Numbers in bold-face type are customary references to sources cited